Small-Scale Grain Raising

Small-Scale Grain Raising

SECOND EDITION

Gene Logsdon

ILLUSTRATIONS BY
Jerry O'Brien

CHELSEA GREEN PUBLISHING
WHITE RIVER JUNCTION, VERMONT

All photographs by Gene Logsdon unless otherwise credited.
All illustrations by Jerry O'Brien, unless otherwise credited.

Project Manager: Emily Foote
Developmental Editor: Ben Watson
Copy Editor: Cannon Labrie
Proofreader: Helen Walden
Designer: Peter Holm, Sterling Hill Productions

Printed in the United States of America
Second Edition. First printing, March, 2009
10 9 8 7 6 5 4 3 2 1   09 10 11 12 13 14

The Chelsea Green Publishing Company is committed to preserving ancient forests and natural resources. We elected to print this title on 30% postconsumer recycled paper, processed chlorine-free. As a result, for this printing, we have saved:

34 Trees (40' tall and 6-8" diameter)
12,531 Gallons of Wastewater
24 million BTUs Total Energy
1,609 Pounds of Solid Waste
3,019 Pounds of Greenhouse Gases

Chelsea Green Publishing made this paper choice because we and our printer, Thomson-Shore, Inc., are members of the Green Press Initiative, a nonprofit program dedicated to supporting authors, publishers, and suppliers in their efforts to reduce their use of fiber obtained from endangered forests. For more information, visit: www.greenpressinitiative.org.

Environmental impact estimates were made using the Environmental Defense Paper Calculator. For more information visit: www.papercalculator.org.

Our Commitment to Green Publishing

Chelsea Green sees publishing as a tool for cultural change and ecological stewardship. We strive to align our book manufacturing practices with our editorial mission and to reduce the impact of our business enterprise on the environment. We print our books and catalogs on chlorine-free recycled paper, using soy-based inks whenever possible. This book may cost slightly more because we use recycled paper, and we hope you'll agree that it's worth it. Chelsea Green is a member of the Green Press Initiative (www.greenpressinitiative.org), a nonprofit coalition of publishers, manufacturers, and authors working to protect the world's endangered forests and conserve natural resources.

Small-Scale Grain Raising, Second Edition was printed on 60-lb. Joy White, a 30-percent postconsumer-waste recycled paper supplied by Thomson-Shore.

Library of Congress Cataloging-in-Publication Data

Logsdon, Gene.
Small-scale grain raising : an organic guide to growing, processing, and using nutritious whole grains for home gardeners and local farmers / Gene Logsdon. – 2nd ed.
p. cm.
Includes index.
ISBN 978-1-60358-077-9
1. Grain. 2. Organic farming. 3. Farms, Small. 4. Cookery (Cereals) I. Title.

SB189.L79 2009
633.1'0484–dc22

2009000820

Chelsea Green Publishing Company
Post Office Box 428
White River Junction, VT 05001
(802) 295-6300
www.chelseagreen.com

CONTENTS

RECIPES

INTRODUCTION TO THE SECOND EDITION

When this book was first published in 1977, I suspected, but could not know for sure, that a day would come when increasing populations and increasing costs of producing and transporting food with fossil fuel, fossil fertilizers, and genetic manipulation would cause food prices to rise so high that more traditional production methods—organic, natural, low-labor, and local—would begin to rule the economy. Thirty years later, that is exactly what is happening. Whenever in history a new, more economical way to do anything is discovered, it will take over the market, no matter how hard entrenched big business and government try to stop it. Not all the forces of power, with their sickening subsidy mentality for the rich, can prevail forever against economic reality.

This book is intended for the pioneers of this new, low-cost way of making food—those gardeners and "garden farmers" or cottage farmers who are interested in increasing both the quantity and quality of their homegrown food supply by incorporating whole grains and dry beans into their fruit and vegetable growing systems. I am not writing this for commercial grain producers, who know far more about their business than I do. I can't even drive one of those huge and complicated tractors that can plow an acre a minute. Nor do I wish to denigrate commercial farmers: Some of them are close friends. While I fear that their way of agriculture cannot ultimately sustain itself, we would be in desperate straits right now without these farmers. I would hope that these larger-scale farmers would find ideas and viewpoints here interesting and perhaps persuasive, especially if they are organic or natural farmers. But the methods I describe and argue in favor of do not promise what the agricultural experts call "top profits," but only good food and the satisfaction of producing it on a scale that society can afford.

We have become a nation dangerously dependent on politically motivated and money-motivated processes for our food, clothing, and shelter. In the world we must live in from now on, to produce our own food is the beginning of independence. To accept that responsibility is the first step toward real freedom.

CHAPTER ONE

Homegrown Grains

The Key to Food Security

I remember the first year we grew grains in our garden. A good gardening buddy dropped by one day early in July just when our wheat was ripe and ready to harvest. He didn't know that though. His reason for stopping was to show me two splendid, juicy tomatoes picked ripe from his garden. After a few ritual brags—and knowing full well that my tomatoes were still green—he asked me in a condescending sort of way what was new in *my* garden. I remembered the patch of ripe wheat. "Oh, nothing much," I answered nonchalantly, "except the pancake patch."

"The *pancake patch?*" he asked incredulously.

"Yeah. Sure. Until you've tasted pancakes fresh from the garden, you haven't lived."

"And where might I find these pancakes growing?" he queried sarcastically, to humor my madness.

"Right up there behind the chicken coop in that little patch of wheat. All you have to do is thresh out a cupful or two, grind the grain in the blender, mix up some batter and into the skillet. Not even Aunt Jemima in all her glory can make pancakes like those."

My friend didn't believe me until I showed him, step by step. We cut off a couple of armloads of wheat stalks, flailed the grain from the heads onto a piece of clean cloth (with a plastic toy ball bat), winnowed the chaff from the grain, ground the grain to flour in the blender, made batter, and fried pancakes. Topped them with real maple syrup. Sweet ecstasy. My friend forgot all about his tomatoes. The next year, he invited me over for grain sorghum cookies, proudly informing me that grain sorghum flour made pastries equal to, if not better than, whole wheat flour. Moreover, grain sorghum was easier to thresh. I had not only made another convert to growing grains in the garden, but one who had quickly taught me something.

Grow Your Own Grains

The reason Americans find it a bit weird to grow small plots or rows of grain in gardens is that they are not used to thinking of grains as food directly derived from plants, the way they view fruits and vegetables. The North American, unlike most of the world's peoples, especially Asians and Africans, thinks grain is something manufactured in a factory somewhere. Flour is to be purchased like automobiles and pianos. Probably this attitude came from the practice of hauling grains to the gristmill in past agrarian times. Without the convenience of small power grinders and blenders of today, overworked housewives of earlier times were only too glad to have hubby haul the grain to the gristmill. And that gave him an excuse to sit around all day at the mill talking to his neighbors.

But even with the advent of convenient kitchen aids to make grain cookery easier, the American resists. He will work hard at the complex task of making wine—seldom with a whole lot of success—but will not grind whole wheat or corn into nutritious meal, a comparatively easy task. I know, because I was that way myself. Until I saw with my own eyes that a good ten-speed blender or kitchen mill could turn grain into flour, I hesitated. Now it boggles my mind to remember that for most of my life I lived right next to acres and acres of amber waves of grain, where combines made the threshing simplicity itself, and yet our family always bought all our meal and flour.

The real tragedy of that ignorance was that the flour we purchased usually was the kind that had been de-germed and de-branned too. Most of the nutrition had been taken out of that flour to give the American home cook what she seemed to want: a pure white powder that would last indefinitely on the shelf and make pastries of fluffy, empty calories.

What has sparked a new, or renewed, interest in homegrown grains is the dramatic rise in grain prices, and rumors of shortages worldwide, that occurred in 2007. Whether these high prices and shortages are the result of ever-rising populations in so-called third world countries, the dramatic increase in the price of oil, or the greater use of corn and other food plants for making biofuels, we can't say for sure. Nor can we predict whether these conditions will continue. But we have been warned. For a whole host of reasons, it is time to think about growing your own bread.

The nutritional picture for whole grains is getting better all the time, thanks to the progress being made by plant geneticists. There are, first of all, the problematical GMO advances (genetically modified grains), which make modern chemical and large-scale farming easier. It is too early to predict what this development will mean for the future. So far, these genetic wonder plants haven't meant bigger yields or haven't produced a farming method that third world (or perhaps even first world) countries can afford. But some of these developments, which can stack disease-attacking genes into grains (or into products like milk from cloned animals) may indeed have medicinal value and justification. It's too soon to know.

But outside the gene-stacking laboratory, dramatic developments in grain quality and production are being achieved. Opaque-2, or high-lysine corn, with almost twice the normal amount of the proteins lysine and tryptophan in it, indicates the possibility of more improvements. Triticale, a cross between wheat and rye that does not always live up to its promises, sometimes outyields wheat, oats, rye, and barley and has more protein than ordinary corn. New varieties of oats, long known as the grain with the highest protein (excluding legume seeds like soybeans), range as high as 17 percent protein content. And the cholesterol-fighting benefits of oats are well established now. Studies of new buckwheat varieties have prompted the USDA's Agricultural Research Service to announce that this traditional crop, which made something of a comeback in the 1990s, has an amino acid composition nutritionally superior to all cereals, including oats. There's also renewed interest in traditional grains like spelt, which a few gluten-intolerant people may sometimes be able to handle in place of wheat. And, perhaps most exciting of all, Wes Jackson's Land Institute in Kansas is developing perennial grain from wild wheatgrasses and crosses with wheatgrass and wheat. Think of what it would mean if we could plant a grain like we would any other grass, and harvest it every year without any planting or soil cultivation needed.

All sorts of projects seeking to develop traditional grains and keep them inviolate from GMO grains are ongoing. The Farmers Breeding Club of the Northern Plains Sustainable Agricultural Society is a project linked to a series of organic-variety trials with small grains being conducted through a partnership between organic growers and agronomists at North Dakota State University and the University of Minnesota (www.npsas.org/BreedingClub.html). In another example, Canadians are bringing old heritage varieties of wheat back into circulation, and using them in bread making (http://members.shaw.ca/oldwheat/).

This is probably as good a place as any to say something I will probably repeat until you get tired of reading it. I have discarded almost all of the general references to sources of grain information that were in the first edition. They were either outdated or too general to be helpful. The best way to stay abreast of new information on grains is to use the search engine of your choice on the Internet. *Everything* is on the Internet. But even better than that is to involve yourself in *local* activities in small-scale farming. There are all sorts of new organizations and efforts in place that amaze me, even though I thought I was more or less in the flow of this information. For example, I was looking for places where a small grain grower could get a small amount of seed cleaned (by and by I will talk about the need for seed cleaning). In earlier days, every farmer had a seed cleaner. Now, hardly anyone does. I was about to write that you would have to take your grain to an elevator or farm-supply service to get it cleaned when I happened to check the membership directory of the Ohio Ecological Food and Farm Association (OEFFA), of which I am a member. To my surprise, not only did one of our members offer seed-cleaning services at his garden farm, but *he lived just a few miles from me in the same county*.

Organizations like OEFFA flourish in nearly every state now, certainly in every geographical region. Home in on them. They all have newsletters about their projects, and these newsletters contain advertising from other garden farmers about the products and services they offer. This is up-to-the-minute information, which no book can promise. My latest favorite "find" is the Northern Plains Sustainable Agriculture Society mentioned above.

Almost all grains can be sprouted to make delicious salads, and in some ways are more nutritious than the dried grain. Beans, clover (especially alfalfa), and wheat make excellent sprouts for human consumption. But oats and barley, in addition to wheat, can be sprouted and fed to chickens and livestock as farmers sometimes do. That kind of feed supplement can keep farm animals healthy and well-fed even in winter without today's expensive all-vitamins-included commercial feed.

Corn sprouts win no prize for taste, but corn makes up for that lack with other advantages. Sweet corn and popcorn are two of our most popular foods, but corn can also be parched, pickled (corn salad), or made into hominy. Popcorn made the national news in 2008 because of the prices being charged for it at movie houses. I found that simply ridiculous. There is nothing easier to grow than popcorn, or easier to prepare for eating. Pioneers in the Corn Belt survived some winters almost entirely on a diet of corn. They cracked, ground, grated, boiled, parched, squeezed, flaked, and baked it into porridges, cakes, muffins, dodgers, and "pone."

A very important food use for grains is in making alcoholic beverages. The best moonshine I ever tasted was "made right" from fermented corn mash. It equaled in mellowness the most expensive firewater I can afford. Of course, other grains make other kinds of whiskey, and malt from barley, a leading crop in the northernmost states, is used for beer and other malt foods and drinks, and of course Scotch whiskey. Wheat beer has also become popular, as has vodka from wheat and other small grains.

Whole Grains for Your Livestock

But the use of whole grains directly in your own diet is only half the reason for growing them. The other half, just as important I think, is to ensure yourself and your family an economical, steady supply of milk, meat, and eggs, and possibly cheese, wool, or other animal products you need or desire as part of your goal of homegrown security. I believe in and practice grass farming or pasture farming, where animals get most of their nourishment from perennial grass and clover pastures. Pasture farming makes a small amount of grain in the animal's ration practical because small-scale farmers simply do not have the wherewithal to raise large amounts of grain even if they wanted to. Pigs and chickens, both of which lack the multiple stomachs of grazing animals like cows, sheep, and goats, especially benefit from some grain in their diets. If you have to go to a store to buy the grains you need for your chickens or pigs, your own home-raised meat and eggs will cost you nearly as much as if you had bought them from the store. Furthermore, if you have to buy your grains in the marketplace, you may have to settle for less nutritional quality than what you could grow on rich organic soil and then air-dry by traditional, natural methods rather than with artificial heat.

Grain plants often give you other important products besides the grain. Wheat and oats, rice and barley give you straw as a by-product—the dried stalks left after the grain is threshed. Straw makes excellent bedding for animals and mulch for the garden. It can be woven into baskets too, and in recent years it has even been much in demand for building straw-bale houses, a traditional form of "green" construction that is enjoying a renewed popularity. Corn leaves dried or silaged are good roughage feed for cows. Corn husks can be plaited into strong rope, fashioned into dolls and decorations, or used to fill a mattress in a pinch. Cane sorghum makes good syrup; buckwheat and clovers provide the bees with an abundant source of pollen for honey making. And, not to be outdone, oats provide the hulls that the manufacturer of Rolls-Royce autos once used to polish the cylinder sleeves of their expensive cars. Maybe they still do.

Cultural Pros and Cons

Finally, the special advantage of grains for the organic gardener and farmer is that you can grow them more easily with organic methods than you can fruits and vegetables. All grains except corn will withstand low fertilization better than vegetables. Field beans, especially soybeans, will add nitrogen in the soil. Corn is easier to cultivate mechanically than fruits and vegetables because it grows well in confined rows, making mechanical weed cultivation easier. Fungal disease is less of a threat in grains than in fruit. Grains have their share of insect enemies, but control is not nearly as critical as it can be in fruits and vegetables.

Dry beans and buckwheat can be planted as late as July 10, except in the far north, so they can be double-cropped behind peas, early beans, lettuce, or strawberries. A late sweet-corn patch may work out well as a second crop too. Barley and wheat can be planted in the fall after other crops are finished and harvested the next summer in time to double-crop that soil to late vegetables.

How Much Grain?

Even a modest harvest of a peck of grain will make a lot of meals—believe me. Excess ears of sweet corn needn't go to waste, either. Dry the corn, shell it, and make cornmeal in the blender. Or parch the corn over the fireplace on a winter evening.

Almost everyone who becomes familiar with the tastiness of whole-grain cookery wants to pursue it. Even if you don't grow your own grains, you'll not find a better way to make your food dollar pay than to buy grains and cook from scratch. And you'll soon find out how much grain you need or want to use for a year. It won't be as much as you think, even if you bake *all* your own bread and pastries.

We bake bread every week, and my wife makes a variety of cookies, cakes, pancakes, shortcakes, pie crusts, and cooked dishes with our whole grains. If the grain is ground fine enough, it makes good bread without the addition of any white flour, though we do add a little because we think it makes the bread a little lighter.

A bushel of wheat makes about fifty 1-pound loaves of bread. Two ears of corn make enough cornmeal for a meal's worth of corn muffins. The grain, as you can see in table 1 on page 8, expands as it cooks with water, and so gives more food to eat than you would think the uncooked grain represented.

At most, figure a year's supply of wheat at 4 pecks (1 bushel); corn, 2 pecks; popcorn, 2 pecks; soybeans, 4 pecks; grain sorghum, 2 pecks; buckwheat, 1 peck; oats, 1 peck; triticale or rye or barley, 1 peck; navy or other soup beans, 2 pecks; alfalfa for sprouting, 1 or 2 quarts; lentils, field peas, cane sorghum (for flour), about 2 quarts each. But only experience can give you the precise annual amounts needed. We don't grow and eat as much as suggested here, but could if we wished, without increasing our production labor noticeably. Of course you can gauge your own family's consumption by estimating how much flour, cornmeal, and other grain products you use now. But your own grains may prove so delicious that you will want more than that.

TABLE 1. COOKING GRAINS

Grain	Amount Uncooked	Amount of Water	Cooking Method[1] and Time	Amount of Cooked Grain
Barley	1 cup	4 cups	boil 30–40 min.	4 cups
Buckwheat	1 cup	2–5 cups	boil 20 min.	3 cups
Cornmeal	1 cup	4–5 cups[2]	double boiler 30–40 min.	4-5 cups
Millet	1 cup	4 cups	boil or double boiler 25–30 min.	4 cups
Oatmeal	1 cup	2 cups	boil 10 min.	4 cups
Rice (Brown)	1 cup	2–2½ cups[3]	boil 35–45 min.	2½ cups
Rye	1 cup	4 cups	boil 1 hour	2 cups
Soybeans	1 cup	3–4 cups	boil 2–3 hours	2½–3 cups
Triticale	1 cup	4 cups	boil 1 hour	2½ cups
Wheat	1 cup	3–4 cups	boil 1 hour	2½ cups
Wild Rice	1 cup	4 cups	boil 40 min.	3–3½ cups

Notes:

1. All the above grains except soybeans may be cooked by the thermos method. Bring required amount of grain and water to a boil, pour into a wide-mouthed thermos, close and leave for 8 to 12 hours.

   Another way to cook grains is the "pilaf" method. This involves sautéing the grain, usually with minced onion, in oil and then adding stock or water, approximately twice as much liquid as grain, and cooking it, covered over medium-low heat until the liquid is absorbed and the grain is tender. The time is about the same as above. Brown rice, barley, millet, and wild rice are especially good cooked this way. Buckwheat is traditionally cooked in this way, but a raw egg is stirred into the dry grains before adding the stock or water. This replaces the need for sautéing the buckwheat in oil, and is done to keep the grains separate throughout the cooking. The required amount of water is 2 cups for the "egg" method of cooking buckwheat, and 5 cups when cooking it to be eaten as a cereal.

   The harder grains such as wheat, rye, and triticale may be brought to a boil in the required amount of water, boiled for 10 minutes, then left to soak for 8 to 12 hours in this same water. After the long soaking, they may be cooked for 15 to 20 minutes and will be tender enough to eat. This is one way to shorten the cooking time.

   The pressure-cooker method offers the advantages of cutting the cooking time in the above chart in half. In general, use twice as much water as grain when cooking in the pressure cooker, although more water—four times the amount of grain—is needed for the harder grains, such as rye, triticale, and wheat.
2. When adding cornmeal to boiling water, it is best to first combine it with 1 cup of cold water and then stir this into the remaining 3 to 4 cups of boiling water. The lesser amount of water is to be used when you wish to have a stiff cooked cornmeal, as for cornmeal mush.
3. The lesser amount of water is required for short- or medium-grain rice, the larger amount for long-grain rice.

A further tip on cooking grains:
To enhance the flavor and shorten cooking time, toast grains in a dry, medium-hot iron skillet, stirring constantly, until they have a pleasant fragrance and take on a darker color. This also enables the grain to be "cracked" or coarsely ground in an electric blender.

Figuring Space Requirements

You don't need much space to raise at least some grains. A normal yield of wheat grown organically would be at least 40 bushels to the acre. So you'd need only 1⁄40th of an acre to produce a bushel. That would be a plot of ground 10 feet wide by about 109 feet long. A really good wheat grower with a little luck could get a bushel from a plot half that size. Wheat yields have been recorded as high as 80 bushels per acre and even higher.

But using the same kind of average calculations as above, table 2, below, shows the amount of space you'd need to grow a bushel of the following grains.

Don't hold me too tightly to these figures. They're estimates to give you an idea of how big the playing field is. Weather, fertility, variety, and know-how could alter these figures. All I'm trying to show really is that 9 bushels of assorted grains might be raised on a quarter of an acre and provide you with the major portion of your diet.

The amount of grain necessary to support a few head of livestock is not large, either. You need about 12 bushels of corn to fatten a feeder pig to butchering weight. We don't feed sheep any grain because we sell lambs fed exclusively on grass and mother's milk. A hen needs about a bushel a year, but if she has ample free range, she needs hardly half that and in a pinch perhaps none at all. A milk cow, along with hay and pasture, needs perhaps 5 or 6 bushels; a beef steer, about the same. And we have raised tasty beef

TABLE 2. GROWING GRAIN BY THE BUSHEL

field corn:	10 feet by 50 feet
sweet corn:	10 feet by 80 feet
popcorn*:	10 feet by 80 feet
oats:	10 feet by 62 feet
barley:	10 feet by 87 feet
rye:	10 feet by 145 feet
buckwheat:	10 feet by 130 feet
grain sorghum:	10 feet by 60 feet
wheat:	10 feet by 109 feet

*for the larger-eared varieties; I don't know per-acre yields for the smaller varieties, like strawberry popcorn.

without any grain. In other words, an acre of corn could fill the grain requirements for one pig, one milk cow, one beef steer, and thirty chickens.

What *is* necessary to raise grains successfully in the large garden or on the small farm is an understanding of planting, harvesting, and processing methods that are no longer common in commercial farming. In many instances, the right way in commercial grain farming today won't be the right way for small homestead growers. In some instances, the right way for you requires use of the latest technologies; in other cases it requires a reaching back for knowledge now almost lost. It takes both to make grain growing and grain eating the cottage industry it once was, and the key to food security it must become if personal independence is to be maintained and personal freedom preserved.

CHAPTER TWO

Corn

America's Amazing Maize

Some of what I wrote about corn in this book's first edition now strikes me as overloaded with the kind of general information that makes a writer sound like an expert, while not being very helpful. In rural Kentucky, where my wife grew up, an expert is defined as someone who can purse his lips up close to the rear end of a mule and in one huff, blow the bit out of its mouth.

Now with thirty more years of experience in growing corn on a small scale, I am going to describe first how we do it. Not that our way is the best way or the only way, but in recounting not only how we do it but why, the reader will hopefully get the information needed without all those general factoids that can make a book heavy enough to serve as a doorstop.

Our way of growing corn on a small scale evolved out of an embarrassment. I became convinced that growing the stuff the "modern" way is environmentally destructive and unnecessarily expensive, but I couldn't quit. I was corn-addicted. Besides, I wanted to raise farm animals mostly on pasture, with only a little grain in their diets and only a little machinery and fossil fuel spent in annual cultivation. Then it occurred to me that a pasture-based farming system suddenly made small-scale corn growing eminently practical. In other words, I could continue my addiction to growing corn by doing it in a way that was not environmentally destructive.

Corn is my grain of first choice for all purposes because, first of all, it is tough stuff. It will survive adversity better than other grains. Also it can be planted and harvested on a few acres with mostly hand labor. Ears of corn are much larger than grains of wheat, oats, barley, rye, or the other "small grains" as they are called. A hundred years ago, it was not uncommon for a farmer to plant and harvest ten to fifteen acres of corn with hand planters or horse planters and hand husking. So one acre, which is about all I need on my small pasture farm, is easy enough to produce. I put two one-row garden planters together to make a two-row planter. Pushed by any healthy human from age fourteen to seventy or older, it's possible to plant half an acre a day easily enough. Our whole family—grandchildren included—harvests the field corn in a couple of weekends.

I grow open-pollinated corn because, as even my friends say, I am a contrary bastard. (There are many of us. A good place to learn more about OP corn is the Michael Fields Agricultural Institute, P.O. Box 990, East Troy, Wisconsin, 53120, www.michaelfields aginst.org). Open-pollinated corn has its drawbacks—usually less yield per acre and a tendency to lodge (blow over) more than hybrid corn. But we think the cornmeal tastes better than that from hybrid corn varieties. This is not just our bias. Acquaintances from across the county go out of their way to buy our corn because they too think the taste is better. Deer and raccoons heartily agree, and those are the two worst problems in raising corn (see below). Invariably, deer will bypass neighboring fields of hybrid corn to get to our corn.

Another reason I favor open-pollinated corn is that our favorite variety ('Reid's Yellow Dent') makes ears that are occasionally a foot long and sometimes longer. Hybrid ears usually average 6 to 8 inches long. Ear size is critical in hand harvesting. I can shuck one big ear faster than two smallish ears.

But the main reason I have settled on this "old-fashioned" corn is that I can save my own seed. You can't save seed from hybrids because the seed will revert back to the parents of the hybrid strains, and this is at least one case where the parents aren't nearly as well-bred as the children. The farmer growing hybrid corn must go back to the seed dealer every year for seed or hybridize his own crosses, which requires expertise most of us don't have. I do not want to depend on agribusiness companies that way, especially when the seed cost soars to over a hundred bucks a bushel, as it has been doing lately. Saving my own seed is my version of hybridizing: it allows me to keep on selecting for greater ear size, greater stalk strength, better taste and softer kernels for the animals to chew and digest. (I usually feed them whole corn, not milled corn.) After some thirty years, I have improved our corn in all four respects, although it still lodges in windstorms more than hybrid corn does. But, since I harvest by hand, I can still

By growing open-pollinated varieties of corn, it is possible to save your own seed from the largest, best ears for replanting, in the same way that farmers have traditionally improved corn varieties over the centuries.

get the ears off lodged stalks, which is not so true of mechanical harvesters.

Another drawback of our OP corn is that it does not germinate as well as hybrid corn until the soil is thoroughly warmed up in the spring. But that is actually an advantage because early planting is risky and encourages weeds to grow faster than the corn, making early weeding more difficult. The small producer does not need to be in a hurry and risk planting corn until the soil temperature has warmed up to at least 65° heading for 70°.

Don't expect open-pollinated varieties to reproduce "true" to the plant from which you saved the seed. All sorts of little variations in ear size, kernel shape, color, ease of shelling, chewability, and maturity dates will come from saved seed. This is part of the fun of it, for me. If the seed came exactly true, as in cloning, there would be no chance for improvement. But the process of selection is slow going. Remember that corn started out many centuries ago as a grain no bigger than a head of wheat. So every year I watch closely for "perfect" ears, that is, ears over a foot long with twenty-two or more rows of kernels; with the rows of kernels slightly separated from each other on the cob for better drying; with no sign of fungal diseases; with stalks that do not fall over in strong winds; and with ears that turn down on the stalk as they mature so that rainwater can't get into them and that then are easily husked by hand. Saving out these best ears every year is a keenly enjoyable pastime because of the potential it suggests. If one had an acre of "perfect" ears, with a per-acre plant population of 20,000 stalks (commercial growers often aim for 30,000 plants per acre, but the small-scale grower with open-pollinated corn should be satisfied with a population of 18,000 to 20,000 plants), each plant bearing one ear of corn with twenty-four rows of kernels on foot-long cobs, the weight of the kernels would average over 1 pound per plant. (I know; I've weighed them.) The yield per acre would then be over 20,000 pounds of grain. Figuring shelled corn at 60 pounds per bushel, that would mean over 333 bushels per acre, beating the world record for hybrid corn. A good yield in commercial cornfields today is 200 bushels per acre. On most soils, 180 bushels per acre is commendable. Open-pollinated corn usually yields around 110 bushels per acre presently and sometimes it's not that good. But if more growers got interested in saving seed from perfect ears, that could change to a higher number over time. Look how many centuries it took the Mayans and their forerunners to get ears up to 6 inches long.

Obviously, after yield considerations, you next want to decide how much land you need to raise your corn. To make the math easy and to take into account land of only average fertility, let's figure 100 bushels of corn per acre, remembering that the actual yield could be twice that or something a little less than that. We saw in chapter 1 that a fertile acre can produce enough corn for a pig, a milk cow, a beef steer, and thirty chickens. If you are operating mainly a pasture farm where grass and clover for grazing and hay are the main animal feeds, then that one acre is all you need for animal feed and your own sweet corn, popcorn, and other assorted decorative or specialty corns for your table. Some animals don't need any grain. We raise about twenty to thirty lambs from a flock of twenty ewes every year on about fourteen acres of rotated pasture with no grain at all, but they do eat the corn fodder. I feed our dozen laying hens three or four ears of corn a day when they are getting plenty to eat eight months a year from foraging in the woods and pasture. I give them a little more whole grain in winter. We fatten about thirty broilers a year, and they do get regular milled grain along with our own whole grain because I want them to fatten in six weeks and get them processed and into the freezer and out of harm's way. This is more to suit my writing schedule and to avoid hawks than our homestead schedule. I could raise them on whole grains and free-range grazing of grass and insects, but it would take longer, and these heavy fat chickens would be easy prey for foxes, raccoons, and other predators.

We have raised beef on good clover pasture and mother's milk with no grain, but where absolutely first-rate clover pastures are not available, you will usually want to feed some corn after a calf gets beyond 300 pounds. A pig needs about 12 bushels of corn to fatten to 200 pounds, along with good clover hay or pasture. Nor does the corn have to be milled, especially with softer, open-pollinated corn. The old bible of livestock feeding, Morrison's *Feeds and Feeding*, says that the first hundred pounds of a pig's weight can be produced feeding ear corn alone. With good pasture or hay, I think all a pig's grain could be from whole corn. You can feed ear corn to steers too, especially if you break the ears into shorter pieces. Slap an ear of corn sharply over the edge of a board, and it will snap into two pieces readily. In feeding ear corn to chickens I usually shell the kernels off the ear although the chickens learn to do it quite well themselves. To shell a couple of ears at a feeding, I often rap them sharply against a board or a stump. The kernels come flying off.

I could go on and give detailed and expert formulas for how much and in what portions you should feed corn with other grains to animals. I won't, though, because most of that kind of advice is mere marketing palaver put out to sell commercial feeds or to sell more grain than an animal on free range needs, or to help the commercial producer of meat and milk gain the absolute nth degree of so-called efficiency (see chapter 12). The small-scale grain grower, with good pasture, can ignore most of that kind of information and rely on common sense and experience. You will be feeding your animals table scraps and surplus garden vegetables along with grass and clover (don't overlook lawn clippings) and only enough grain to make up the difference. You will not be interested in trying for the absolute highest rate of gain or the absolutely highest yield. The "right" way in large-scale commercial agriculture is not necessarily the right way for the homestead gardener and farmer.

Soil preparation is the next consideration. Work the soil for field corn about like you do for sweet corn in the garden. We plow under a green-manure crop (old hay stand) in early spring, then let the plowed-up soil settle at least a couple of weeks, then disk the land twice and then go over it with the garden tiller. Because we grow the corn in four-row strips rather than a solid field, we would be better off to use a heavy tractor-mounted rotary tiller for the whole cultivation process, avoiding the dead furrows that come with plowing, but that's more expense that so far I have managed to do without. Corn needs to have a finely worked seedbed for good germination, unless you have a no-till planter, which is usually beyond the practical affordability of the small-scale grain raiser. And no-till planting usually requires using herbicides to kill off competing weeds.

Planting in strips is not necessary but a personal preference of mine. In a corn field, the outside rows always yield better, and so with strips rather than a solid plot there are more outside rows. Also, the soil in the strips is not as shaded as is the soil in a plot planted solidly to corn, so when I sow clover into the standing corn after weed cultivation for future pasture or hay, or sow rye for pasture or grain in the following year, these crops grow better. I have also inter-seeded oats into standing corn in July for winter grazing, and the oats need all the sunlight they can get. Also, if you are planting in a sloping field, the strips can often reduce erosion significantly compared to a plot entirely of cultivated soil.

I try to plant the main crop of corn in the first week of May here in northern Ohio. Sometimes weather doesn't cooperate, but almost always, I have all the corn planted by May 21 except the late-succession plantings of sweet corn. I use two hand-pushed garden planters connected by a homemade frame (three 1 × 3-inch boards) with space enough between them for the garden tiller to operate in cultivating weeds. Farmers argue forever about the right depth to plant corn, anywhere from ¾ inch to a little over 2 inches deep. If you are planting very early, before the soil has warmed up to at least 65°F but moist conditions prevail, just barely cover the corn so that it gets more benefit from whatever warm sun is available. If the soil is warm and moist and you are going into your regular time for planting, figure no more than 1 inch. If the soil is warm but dry, 1½ to 2 inches might be better. If you could be sure it was going to rain the next day, shallower planting is better. If you could be sure rain was not going to fall for a week or more and the soil is already dry, you might go as deep as 2¼ inches. But you never know for sure, so you try to hit a happy medium, or go on gut instinct, and when you guess wrong, beat your head like I do. Corn is a very resourceful plant. Most of the time, if you don't get into a big hurry and plant before the soil is warm, it will produce well, even if you screw up a little. Leave an ear of corn on bare soil in the fall, and if a deer or wild turkey or one of your farm animals doesn't eat it, in spring some of the kernels on the cob will sprout and grow right there without any soil preparation at all.

The worst mistake you can make, and all of us have made it, is to rush out there and try to cultivate soil when it is not dry enough yet. In heavy clay soils, if you do that you will have to fight hard clay marbles and cue balls all summer. The point is that when you are a small-scale grain raiser, what's your hurry? It took me a while to learn that. Unlike the commercial corn grower looking to get a couple thousand acres planted before May 10, I have at most an acre or two to plant. I've known of corn planted on June 16 that still made 100 bushels per acre. So take it easy. Spring is a nice time of the year. Go pick some wildflowers. Or hunt for morel mushrooms. Stay off the land until beech-tree leaves are as big as squirrel ears, tradition says, and that's about right.

Control weeds as you do with sweet corn in the garden. I don't like to use herbicides, so I weed with the garden tiller and hoe. I roll the loose soil stirred up by the tiller into the rows to bury weeds where the tiller can't reach. Some shovel cultivators can be handily adjusted to do the

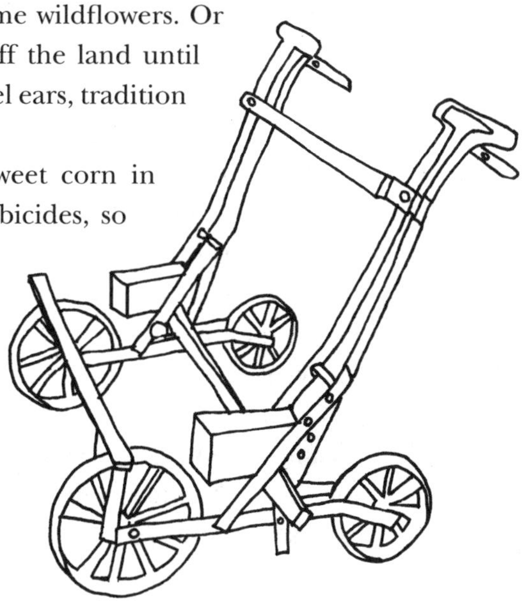

A double hand-push planter.

rolling, but I prefer to shuffle along and roll in the dirt with my feet and the hoe. Hard on shoes, but the work is not too arduous for a half acre or less. You should cultivate at least twice, and three times is better, before corn is knee-high.

Before herbicides, farmers fought weeds with what is called a rotary hoe. Just as the corn appears above the soil, or even before it appears, they would (some still do) pull this implement over the field with a tractor at a fairly high rate of speed. The whirling iron hooks jerk tiny germinating weeds out of the ground without hurting the corn. This is a viable method for organic growers with comparatively large fields of corn. In the garden you mimic the rotary hoe when you go over ground lightly with a rake before your planted seeds come up.

After the corn gets knee-high (too tall to cultivate mechanically) I still sometimes walk through it with a hoe to chop out some weeds, but if the corn is above the weeds it will usually keep the upper hand and shade out some of the competition. Some husbandmen still turn lambs into standing corn in August to eat the weeds and lower leaves of the corn. Since lambs will seldom bother the ears, this is a good economical practice, but I keep my lambs with the ewes (do not wean them) until they go to market. Ewes are sometimes smarter than they look and will knock the stalks over to get to the ears.

Since we are talking here of small parcels of corn, grazing off the lower leaves with lambs, or grazing all of the stalks and ears with hogs or cattle in the fall, which are viable traditional practices, is not within the purview of this book, but you should be aware of these practices. (I do describe them in some detail in my book on pasture farming, *All Flesh Is Grass*, published by Swallow Press at Ohio University Press in 2004.)

As the corn grows tall in August I often cut out nonproducing stalks and stalks with only nubbins on them with a corn knife and feed them to the sheep. You get a few of such stalks when planting open-pollinated corn. Old-timers used to cut the green stalks above the ears after the ears have been pollinated and feed this green matter to livestock too when other feed is short. Pollination occurs as soon as the ears silk out and pollen from the tassels falls on them. Cutting off the tops of the stalks after pollination has the added advantage of reducing the height of open-pollinated corn so it doesn't blow over in storms. It also puts the remaining vigor of stalk growth into the ears. Decapitating stalks with a corn knife is hard work, so I've done it only occasionally. But doing a couple of armloads a day is not so bad. The ears continue to develop just fine without the top-story stalk.

Harvest sweet corn when it is tender and sweet, of course. Field corns are ready to harvest when the kernels are fully dented, dry, and hard. The stalks will have turned mostly brown. You can let the ears hang there on the stalks well into winter or even until spring if you can't get them harvested sooner. Wild animals and birds might eat some of them, however. We harvest in pleasant October weather on weekends convenient for the family to be here. I give the grandchildren a dollar for every ear they find 14 or more inches long. This makes a game out of what might otherwise be boring work for them.

Corn is rarely completely dry even when you harvest it late in the fall. Commercial corn growers almost always dry it artificially, with natural gas as a fuel. This is a very big expense and in my opinion a poor use of fossil fuel. Before corn was harvested as shelled corn, the whole ears were stored in slatted cribs no more than 4 feet wide, and air-dried naturally. That is how I do it. Once corn is cribbed, it can be used as needed, although before you mill it, if you do mill it, give it a month or more in the crib to dry completely. It may look at harvesttime to be completely dry, but it hardly ever is. Corn that we are going to use for our own cornmeal I will hang as individual ears in the garage for quicker and more complete drying.

It might be helpful to list the different kinds of corn and various varieties of all of them, although I doubt it. It is like listing all the makes and models of automobiles. Which seed company sells the best corn is the same as arguing which car company sells the best car. What you choose is a matter of preference most of the time and what you can afford. And sometimes you might want to try a corn (or a car) not particularly recommended by the experts, just to see for yourself. That should be part of the joy of living.

Because corn is such an important part of our economy, plant breeders have put great effort over the years into breeding varieties that develop amazing yields, remarkable disease and insect resistance, increased nutritional value, and now genetically modified strains. If the same amount of effort had been put into pasture plants, we wouldn't need so much commercial corn today, but, alas, that is another subject, fraught with emotion. The most common commercial corn is the standard yellow dent hybrid, which is capable of producing consistent yields of over 150 bushels per acre and as high as 300 bushels per acre. Next in commercial importance are the white hybrids of the southern and central states. Some open-pollinated white and yellow dent varieties are still grown commercially. Specialty corns include waxy maize, a kind of corn first grown for its starch. During World War II, when U.S. supplies of root starches, especially tapioca, were cut off, corn breeders developed waxy maize as a substitute. Waxy corns seem to have feeding advantage in that they are more easily digestible. But like all the specialty corns, yield so far is not high enough to compete much with regular corn.

High-lysine or high-protein corn has been touted as one of the most significant achievements in the fight against world hunger, but, like waxy corns, it must be isolated from other corns to prevent cross-pollination. Yields of high-protein corns have generally been lower than those of regular hybrids, but the higher feed value makes up for the lower yield to some extent. For the average farmer, however, more is better, and so plantings of high-protein corns have mostly waned, not waxed.

Seed-corn companies have successfully crossed high-lysine corn with waxy maize. In tests, the more easily digestible starch of waxy maize combined with the higher protein of high-lysine corn makes for an excellent feed grain for hogs. There's no reason why it shouldn't be an outstanding food grain for humans, too. The kind of starch in waxy maize contains much more simple sugar glucose than regular yellow dent corn, and the proteins in high-lysine are not only better quantitatively, but qualitatively too. But, once again, these crosses do not produce the high yields of regular corn, and so progress in developing them is slow. Mostly you just need to know about them and keep an eye out; it could be their time has just not yet come.

Hybrid field-corn varieties may be simple crossbred hybrids, or single-cross hybrids, or three-way crosses. Some mature in as early as 85 days, some in as late as 120 days, but most are in the 90- to 110-day range. The farther north, the earlier maturing your corn

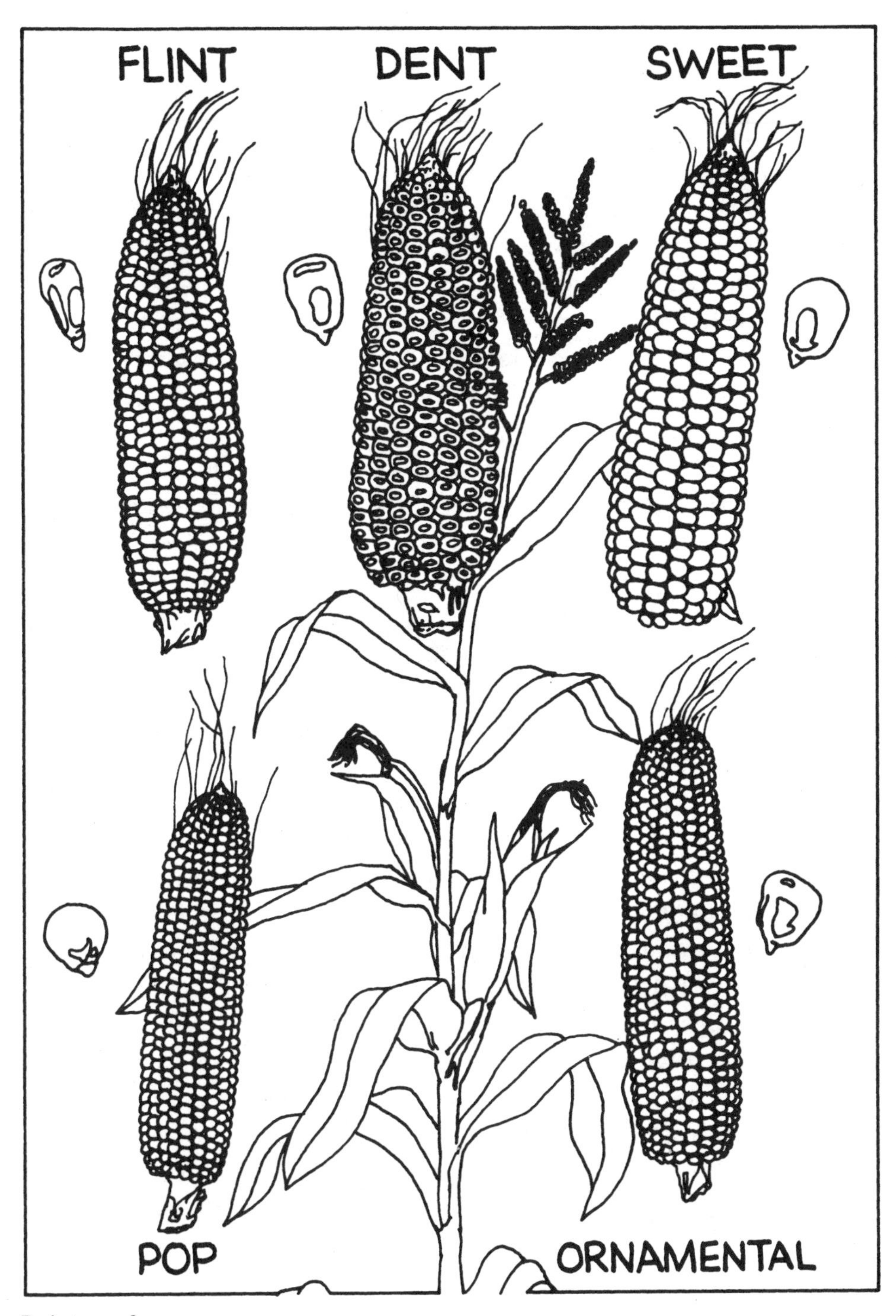

Basic types of corn.

must be to beat the frost dates. Longer-maturing hybrids invariably yield more, as is true with most any crop. The old rule of thumb prevails: plant the varieties that the seed companies sell for your geographical region. In normal years you won't have any trouble finding seed, although as I revise this book in 2008, seed corn is in short supply because this is anything but a normal year in farming. I'm beginning to think every tenth farmer in the Corn Belt is a seed-corn salesman on the side, since so many of them have signs by their mailboxes announcing the varieties of corn they sell. You can buy from these farmers. Or you can buy seed from farm-supply and feed stores or by mail order.

The choice of commercial varieties of open-pollinated field corn is much more limited, but some mail-order seed companies still carry them, and a few farmers all over raise and sell them. Shumway Seeds, an old seed supplier, sells both white and yellow varieties and seed corn for silage varieties. And remember, you can find everything on the Internet.

White hybrid corn is grown in the South and the southern Corn Belt. Oddly enough, a lot of white corn *seed* is grown in the north, near Lake Erie, by the Schlessman Seed Company of Milan, Ohio. The proximity to the lake allows the longer frost-free growing season that the corn needs, while the northern latitude does not have the disease problems that would affect a seed-growing operation in the South. As is true of yellow hybrids, varieties are bred for various climates, from the southwestern states to the central Corn Belt.

At least one open-pollinated white variety, 'Silver King', is available from Gurney Seed & Nursery Company (Yankton, South Dakota, 57078). The catalog says the variety is adapted to as far north as central South Dakota. White corn grows much taller than yellow corn, especially the non-hybrid varieties, with ears occasionally 6 feet from the ground, not the corn a short man would want to harvest by hand!

Types of Corn

Flint Corn

Flint corn is a yellow, open-pollinated corn with a long and skinny cob. Unlike dent, flint-corn kernels do not become dented on top when mature. Flint corn contains little soft starch and does not shrink when dried. The kernels are harder than dent corn, and not as easily digested. Ears of the 'Longfellow' variety (sold by Shumway) are very long and skinny, fun to grow for its unusualness, but hard to shell. Gurney Seeds sells another yellow flint called 'Minnesota 13'. Gurney also sells a 'Rainbow Flint' that is similar to Indian corn, itself a flint corn. Flint corn used to be called Yankee corn, and is better acclimated to colder short-season growing regions. Most flints mature in 90 days or less. Among the myriad varieties of traditional and "Indian" corns that are still around, one extra-early type that has received particular attention lately is 'Roy's Calais' flint corn from Vermont, available from High Mowing Seeds.

Silage Corn

There are corn varieties both old and new that are grown specifically for silage. Silage is green corn chopped and packed in silos or giant plastic bags. In pasture farming, silage in big plastic bags is becoming increasingly important in livestock farming, especially dairy operations.

Sweet Corn

Generations ago, many farmers ate field corn instead of sweet corn. I remember eating some when I was a child. But sweet corns, especially the newer varieties, are so far superior for the human palate that I doubt anyone still prefers field corn. It is just good to know that you can roast any corn to keep from going hungry.

I don't know of any vegetable or grain that has been improved in taste as much as sweet corn over the past fifteen years. And yet, wonder of the free market, you can still get seed of very old varieties if you really do prefer them: Seed catalogs are full of yellow or white or mixed yellow-and-white-kerneled corn, even blue-kerneled, hybrid or open-pollinated, early or late maturing, and varieties with higher sugar content. I am not going to name names because new varieties come out all the time. There are enough to satisfy any taste. Most catalogs still carry the old open-pollinated varieties like 'Yellow Bantam'. One thing in their favor is their aroma when roasted. That old roasted-corn smell rarely lingers in the new hybrids.

About the only difference in the culture of growing sweet corn as compared to field corn—in practical terms anyway—is that the new, ultra-delicious varieties of sweet corn seem more fragile to adverse weather and soil conditions. They don't germinate in cool soils as well as hybrid field corns, and tend to blow over in storms more than hybrid field corn. I have occasionally tied sweet corn to wires strung above the rows to prop them back up after blowing over. 'Silver Queen', a tall, white hybrid sweet corn, is especially susceptible to lodging, but I like the taste so well I keep on growing it.

In selecting sweet-corn varieties for home use, include an early, a midseason, and a late-maturing corn. You want a little corn throughout the whole season, rather than a whole bunch at one time. Plant a 70-day corn early (when the soil has warmed to at least 55° to 60°F). Wait 5 to 10 days, and plant some early midseason corn. Wait another week and plant some late-midseason crop. Wait another week and plant a late-season variety. If you make your first planting in the first week of May, your last will be about the first week of June. The four plantings will then ripen more or less continuously into late summer. If you plant early, midseason and late varieties all at the same time, they will not follow each other as well as planting them a week apart, but will gang up on you and overlap their ripening times. As a normal rule, corn planted early in the year will take a little longer to ripen than the maturity date stated on the packet. Planted late in the season, the corn will mature sooner than the stated time.

Cross-pollination is another thing to keep in mind. In a planting of OP corn, you want the plants to cross-pollinate among themselves, of course. However, you should try to keep your corn at least a couple hundred feet from neighboring fields of hybrid corn. Since corn pollen can carry a great distance on the wind, you may still get some cross-pollination, but with a reasonable setback it isn't as likely. Since my neighbors alternate their fields from corn to soybeans to wheat, I try to plant my plots next to their fields when they are not in corn. I haven't noticed any serious problems with outcrossing in thirty years of farming; it might even be that a little cross-pollination from my neighbors' fields helps to give my OP corn some stronger stalk strength.

However, you should never plant sweet corn in the same field as your OP field corn. The two types will cross-pollinate readily with one another, and you will get ears of field corn that have kernels of sweet corn on the cob (easy to tell, because the sweet corn kernels look shrunken when dry). The best way to avoid this problem if you have to plant the two corns side by side is to plant a late-maturing sweet corn (most of the white varieties) late in the season, with the OP field corn planted earlier in the season. But I wouldn't risk it.

The earliest sweet corns are usually smaller-eared than average and usually yellow varieties. The early to midseason varieties include the combination white and yellow corns. Midseason varieties are the larger-eared, yellow corns. White corns are almost always late-season corn. If you want an opinionated assessment of the four kinds of corn, here's mine. The extra-early corn is fine for that first taste of the season, but it doesn't have the flavor of the others. The combination white and yellow kerneled varieties are very good in taste. The yellow midseason varieties are excellent and yield better than earlier varieties. They are also better for canning or freezing. White sweet corns do not taste as good to me, but coming at the end of the season, when you are tiring of the taste of yellow corns, you will enjoy the change. Ears are bigger and kernels deeper too, so production is greater.

Be sure to pay attention to what the garden catalogs say about cross-pollination of the new supersweet varieties. Don't plant different varieties any closer to each other than recommended.

Parched Corn

Any sweet corn can be parched, making a snack food as good as any in the grocery store. Just leave the ears on the stalk until late fall, then string in the garage or other protected place until completely dry. Then pretend you are popping corn. Put the dry mature kernels in a pan or frying pan with a tiny bit of oil and keep shuffling the pan on the burner so the corn doesn't burn. It's a good idea to put a lid on the pan or the kernels may pop out like popcorn. Salt to your taste. White or yellow sweet corn makes good parched corn. Some seed companies sell varieties strictly for parching.

Popcorn

There is as much variety in size and kernel color in popcorn as in any other kind of corn. You can buy black, blue, red, burgundy, baby yellow, baby white—all of which have their devotees. I prefer regular old yellow popcorn, but who cares? The important thing to remember is that according to the *Journal of the American Dietetic Association* in its May 2008 issue, popcorn eaters get a 22 percent higher intake of fiber than those who don't eat popcorn.

Popcorn is the easiest type of corn to grow. It germinates and comes up faster. We grow an old open-pollinated kind that friends gave us. It doesn't have a name that I know of. It grows as well as hybrid varieties. Popcorns typically require from 95 to 120 days to mature, but I suppose there are exceptions to that. Plant popcorn at the same time you'd plant your field corn or right after your early sweet corn.

Ornamental Corns

All seed houses sell ornamental or "Indian" corn seed. The varieties are usually non-hybrids, so you can save the seed. One exception is 'Purple Husk Hybrid', an Indian corn with purple rather than white husks. New hybrids and new introductions of old open-pollinated varieties are always coming and going. Study your catalogs. Ornamental corn needn't be "wasted" on decoration alone. When you no longer want it for the centerpiece on the table, you can feed it to chickens or livestock.

The Economics of Small Corn Plantings

An accepted principle of agricultural economics states that the more corn (or any grain) produced with a given amount of time and labor, the lower the per-unit cost of production. The more acres farmed, the more acres to spread the cost of farming over. The larger the operation, therefore, the more "efficient" it can be.

The soundness of this principle is no longer taken for granted.

There seem to be limits to the so-called economies of size, above which bigger means more cost per unit, not less. Ask New York City's managers, or the administrators of any large high school. In agriculture, economists are now talking about "optimum-sized" farms, above which per-unit cost tends to rise rather than decline. The optimum-sized corn farm today is around two thousand acres, according to the experts.

But the economic tunes the commercial farmer must dance to have a far different beat than the songs the homesteader sings. In commercial farming the cost of growing corn has soared to around $600 per acre in 2008, and perhaps even more if you include a charge to management the way other businesses figure. Farmland rents run over $250 an acre in some cases. Farmland prices have now skyrocketed to over $6,000 an acre. Fertilizer and artificial-drying costs tripled in 2008 as fuel prices soared.

Obviously, the small homesteader or gardener growing small amounts of grain is not affected nearly as much by this reckless economy. I consider my corn growing to be recreation, mostly, so I can hardly charge a labor or management cost to it. I use my own livestock manure for fertilizer. My fuel cost for an acre of corn is very small because I use my own labor whenever I can instead of fossil fuel. I can raise the grain I need so much cheaper per acre than the commercial farmer that it seems clear to me that commercial farming as practiced today is a dinosaur on its way to oblivion unless it is subsidized heavily.

Of these above-mentioned costs, the small-scale grower can get by for around $90 per acre or less. More than likely he will not get the 160-plus bushels-per-acre average that the $600 cost is supposed to "buy." However, the necessity of getting top yield is not as critical for the garden farmer as it is for the commercial farmer. A homesteader with one acre of corn may be quite content with a 90-bushel yield; that may be all the corn he wants. Whereas the commercial grower, under the onus of $600 cost per acre, might find a 90-bushel-per-acre yield catastrophic.

The farm activities that may mean ulcers for the commercial farmer can spell relaxation for the garden farmer. For that very reason, it is questionable whether labor should be figured as a cost to the latter at all. Gardening and homesteading are two of the very few pastimes (I hesitate to call them "leisure" activities) that can actually save you money. In fact, compared to money spent on golf, traveling, skiing, or flying, raising your own food can be profitable even if you have a crop failure.

I emphasize the noncommercial aspects of home grain production for a particular reason. Some of the methods and ideas I suggest would be considered wrong by a commercial farmer. Some farmers will spray insecticides on corn even when it is not evident that the bugs are hurting the crop very much. This is called "protecting the investment in the crop," and I can hardly blame them for doing so. For the same reason, commercial growers now feel forced, or are in fact forced by their creditors, to take out crop insurance. Certainly a homesteader, whether organic or not, does not have to resort to such stratagems. It reminds me of the time a neighbor accidentally let his herbicide spray drift over the fence and kill my apricot trees. He felt almost as bad as I did, and offered to pay twice what the trees were worth. I don't think he quite understood that money was not the issue. I wanted apricots, not cash. You can't eat cash. And three years of my life were invested in those trees. What is that worth in cash?

It is at least sad, if not alarming, that commercial agriculture has become so straitjacketed by economics. "It's no fun to farm anymore," my neighbor sighed recently. "It's all business and banking and being efficient twenty-four hours a day. Inhuman." Time was when a fellow who was willing to accept a somewhat lower standard of living (by urban definition) could raise a family happily on the farm even if he were a somewhat mediocre businessman. Not anymore. He can't afford to farm "inefficiently," even if he is willing to accept poverty as a tradeoff. Economics will brook no mediocrity and few mistakes. And the man who "succeeds" finds himself on a treadmill. He must always keep on expanding, keep on borrowing money, and live with the risks and gambles of perpetual debt. For each pound of gold, economics demands a pound of flesh.

The "laws" of economics have placed a burden on agriculture that the latter's natural processes can't cope with. Economics assumes that agriculture responds to financial matters in the same way that the more controlled types of manufacturing respond. Economics assumes that a "good" farmer on "good" land will make enough money above operating costs to pay a 6 to 10 percent return on investment, plus a good living wage for the people who do the work on that farm, plus a nice return to management for the farmer-owner-manager. The successful farmer today is expected to pay about 8 percent for the money he borrows, pay his hired help a salary competitive with industry, and have enough money left over to live in a fine house, drive an expensive car, and vacation in Florida every winter.

I have never in the last fifty years found a farm that can do all that. The fact that farms grow steadily larger every year is itself proof that there's never enough profit in the occupation to accomplish these ends. Wherever a farm seems to be economically "successful," there is either money being pumped in from another source or you will find the soil of that farm being gradually depleted of fertility. Nature never heard of money interest. What you take from her you must eventually put back. The farm is *not* a "factory in the field."

The whole underpinning of what seems to be profit in agriculture today is inflation in land values. The land continues to rise dramatically in money value no matter how you farm it, or whether you farm it at all. One farmer told me a couple of years ago that he would have been just as far ahead to let the land lay unfarmed or "rent it to some other sucker" while he worked at a salaried job, then sold the place when he retired. Of course, as we learned in 1982 and again in 1995, farm values, reacting to too much inflation, can occasionally collapse as well. If you have to sell during an economic downturn, you have nothing to show for your work.

The truly profitable farms are those managed for three or four generations under a consistent policy of frugality and conservation, and upon which some high-quality livestock-breeding program or high-value horticultural operation is carried out along with crop production. The small family dairy farm is the best example, but a beef cow and calf operation or a flock of sheep can make it too. The traditional livestock and dairy farms, by their natures, are semi-self-subsistent in the sense that they are self-renewing kinds of agriculture that require the least amount of commercial inputs from outside sources. Most of the feed for the animals is raised on the farm. The operation is labor-intensive, employing the farmer and his family fully in all seasons. Older, smaller machinery is often utilized. Few luxuries are indulged in. Debt is, and has been, avoided like leprosy, and expansion has come only out of savings, not from borrowing. Expansion without borrowing hasn't been difficult, however, because it comes naturally, by an increase in the number of cows in the herd or sheep in the flock. In a good breeding program, this natural increase has had a value equal to the increase in the value of land. Good cows today cost about the same as an acre of farmland, but the cows are raised while more land has to be bought. And the cows in the meantime provide the farm with fertilizer almost for free to substitute for the increasingly high cost of commercial fertilizer.

The livestock farmer, especially when the operation is based more on pasture than annual cultivated crops, can become richer over the years in three ways. First, because of clover in rotations and the return of ample amounts of manure to the land, soil fertility and organic matter increases. Second, frugality puts some money in the bank nearly every year. In fifty to one hundred years this savings amounts to a considerable sum of money upon which to expand without borrowing or to retire on comfortably. Third, as mentioned above, expansion in the herd or flock comes from natural propagation, not money.

Building a rich farm in this way is too slow an undertaking for all but the *real* farming people, of which there have never been enough. You won't often hear about this kind of farming from bankers because they cannot make much money from such an operation. Rarely will you hear it praised by agribusiness either, because one way the traditional livestock farmer can save money is by refraining from the allurements of the biggest tractor and the latest equipment agribusiness would have him buy. But if you have dreams of turning your homestead experience into becoming an independent farmer, this is the model you should follow. You will find it a good life if you enjoy hard work or the satisfaction that comes from it. If you study the history of this kind of farm, you will find that often it started with a homesteader who originally paid for his land out of off-farm earnings, just as you are doing.

TABLE 3. SOME SUGGESTED CROP ROTATIONS

Grains and strawberries

1st year:	Strawberries
2nd year:	Strawberries; plow under after harvest, and plant buckwheat for late crop
3rd year:	Soybeans; plant to wheat in the fall
4th year:	Wheat, sowed to clover in spring
5th year:	Clover, plowed under in spring, plant corn, then sow rye in late summer while corn is still standing
6th year:	Plow under rye, plant strawberries

Grains and vegetables

1st year:	Corn; sow rye in late summer while corn is still standing
2nd year:	Plow under rye, plant peas, double crop to late fall cole vegetables
3rd year:	Tomatoes
4th year:	Beans: string, lima, dry beans; plant wheat in the fall
5th year:	Wheat; sow clover in spring
6th year:	Clover planted under; plant corn

Grains and hay for livestock

1st year:	Corn, sow half to barley in fall after corn harvest
2nd year:	Sow oats on other half; plant soybeans after barley harvest in June; plant wheat in fall after oats and soybean harvests
3rd year:	Wheat; sow alfalfa in spring
4th year:	Alfalfa for hay
5th year:	Corn or more hay
6th year:	Corn; plant wheat in fall
7th year:	Wheat; sow alfalfa in spring
8th year:	Alfalfa
9th year:	Alfalfa
10th year:	Corn

These rotations are only a few you can follow. With experience, you'll want to vary them. Notice though, how the ground is used fully, often double-cropped, and hardly ever left completely bare over winter.

Providing the Necessary Fertility

Fertilizing a corn crop is not something you start to think about on the day you plant. Fertility is a year-round and lifelong building program on a farm, its peaks coinciding with the corn crop in your three-, four-, or five-year rotation. You build your rotation around corn because it is the most voracious feeder on soil nutrients, especially nitrogen. Fertilize properly for corn and you'll take care of most of the nutrition for soybeans and small grains in your rotation. Or, if you farm organically, put your manure on your corn ground and build fertility by rotating that land with three years of legume hay or pasture.

Rotation of crops is important, whether you are planning a few rows of corn in the garden or several acres on your homestead, or several fields of a farm. Corn should follow a legume, either beans or clover, on an organic farm. Clover, especially alfalfa, can draw into the soil up to as much as 100 pounds of nitrogen every year, all free from the air. That nitrogen alone will go a long way toward making your corn crop successful.

In gardens or on small plots, rotations can take advantage of double-cropping techniques, as suggested in the rotation plans in table 3 on page 31.

On a pasture farm where almost all feed for animals comes from grass and clover, rotating with grains or other crops is very simple because only a small amount of the land is in grain in any given year. We now grow only corn for grain in strips that move every year in the pasture plots. The strips are always on land that has been in legume hay or pasture for at least the five preceding years.

All plant residues should be carefully worked into the soil either after the harvest directly, or after they have been used for bedding or feed and turned into manure. A green-manure crop every fourth year in rotation, an annual treatment, if possible, of 5 to 10 tons of manure, lime at the rate of ½ ton per acre every five years or as needed, and a couple hundred pounds of chemical phosphorus and potash fertilizer, if needed, plus a legume in regular rotation, and your corn should get all the nourishment it needs. The problem is that sources of mined phosphorus and potash are dwindling, and the price is rising. This makes organic farming with manures more practical than ever. Heavily manured ground will get enough potash and phosphorus, generally speaking. An application of potash rock or greensand will add more potassium to the soil if the soil is high in organic matter and dynamic with microbial life. Often, though, some of the potash in greensand and potash rock remains unavailable to the plant for long periods of time. On my small acreage of corn, I spread the ashes from our woodstove that keeps us warm in winter. Wood ashes contain both potash and lime.

Where more potash and other minor elements are needed than organic growers can get from traditional sources, you might want to investigate special fertilizers natural farmers use and accept as organic fertilizers. Some seaweed and sea-product fertilizers contain potash and trace elements, as do some special mineral and organic blends. The best place to find sources of these fertilizers are the advertisements and articles in the magazine *Acres U.S.A.* Much controversy rages about these soil amendments and fertilizers, and I do not know who is right or wrong. I've never used any of them, and my crops always seem healthy so long as it rains (but not too much).

Planting Somewhat Larger Acreages

To plant more than an acre or so of corn, you are going to need some sort of mechanical help, since hand planting gets to be too slow and tedious for above that size plot. Several hand-pushed mechanical planters are available. For an acre or more you should consider planters that attach to either a garden and lawn tractor or to a larger farm tractor. Any dealer who sells garden tractors should be able to get you a planter, too. Manufacturers have come up with the "unit planter" idea. You can buy one unit or, heavens, as many as your tractor can pull. The planters can be purchased with the very latest in precision-planting equipment for large and small seeds. Prices of farm equipment (anything made out of metal) are on the rise now, so prowl around the lots of machinery dealers and see what you can bargain for. (For more information, see the glossary of grain equipment.)

Remember that if you plant two rows at a time (with a two-row planter) then you can cultivate two rows at a time. But you can't use a two-row cultivator on a field that has been planted with a one-row planter. The two-row planter obviously plants two rows that are *always* the same distance apart. But no one could guide a one-row planter so perfectly as to make twin rows with constant, uniform space between them.

Don't forget to set the row marker when you are planting, so you have a guide to follow for the next row. You must keep your rows as parallel and equidistant from each other as is humanly possible, especially when using a two-row planter to be followed by a two-row cultivator. I think I can illustrate the reason better than I can write it.

Here's the pattern of rows planted with a two-row planter: You start at A:

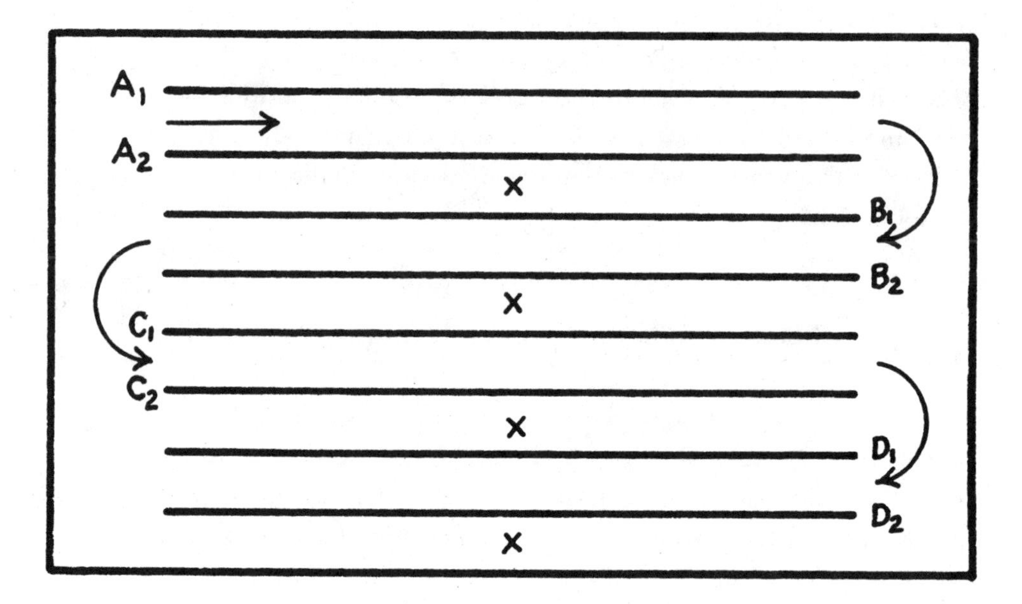

You know that rows A-A are always the same distance apart as are B-B, C-C, and D-D, but the distance *between* the sets of two rows—marked by X—will be the same only if you plant very straight rows and follow the row marker very assiduously. If the X distances are too close, the outside cultivators will cut into the next row. If, for example, A_2 is too close to B_1 the outside cultivating shovel may dig into B_1 as you are cultivating down rows A-A. Nobody can plant the rows perfectly; just don't get them too crooked.

Commercial hybrid corn can be planted as thickly as 30,000 kernels per acre if plenty of fertilizer and moisture are available, but don't you do it. Organically, you'd need at least 25 tons of good manure per acre to support that kind of plant population. Moreover, you'd have to plant the corn in rows not more than 30 inches apart, a move that will facilitate neither hand harvesting nor sowing a cover crop in the corn in late summer.

If you are going to plant an acre of corn without chemical fertilizers, you'll be better off to shoot for a plant population of around 18,000 kernels. That means a spacing of about 8 to 9 inches between stalks in the row, with rows 40 inches apart. Such widely spaced plants will make good use of normal, natural fertility of green-manure crops and manure, and will produce big ears, easy to harvest and process by hand.

If you plan to plant pole limas or pole string beans to climb the cornstalks (saves having to put up poles) the corn plants in the row should be even more widely spaced, one stalk every 12 to 15 inches. When the corn plant is 6 inches high, plant a bean on either side of it about 6 inches away. If moisture is normal, both corn and bean will grow well, the latter fixing a little nitrogen in the soil for the corn to feed on.

In the garden, you will want to plant your sweet corn, popcorn, and decorative corns more thickly than described above. Rows need be only 30 inches apart—even closer if you are going to cultivate by hand—with plants in the row 6 to 8 inches apart. It may be better to plant even thicker than that, then thin out weaker plants. But if you want to grow pole beans up the cornstalks, the same rules apply as described in the preceding paragraph, or else grow the beans only on the outside rows.

Enemies of Corn

Wild animals are now the biggest threat to small-acreage corn growers. We have become an urban nation where most people do not hunt and are in fact adverse to hunting. I have tried unsuccessfully to convince nature lovers, of which I am one, that deer, raccoons, squirrels, rabbits, wild turkeys, and Canada geese, to name a few, do not need protection, but my words fall on deaf ears. The only thing that will cure this point of view is for the person holding it to take up gardening. Then he or she will learn. It amuses me that wild animal defenders feel no compunction about killing moles in their yards or rats in their cellars, but do not understand that the same logic must influence our attitude toward any animal that overpopulates in human habitats and does not have its own natural predators anymore.

So I will not discuss the philosophy of animal control in corn because those who produce food already agree with me, and those who believe all wild animals need protection are not going to listen. I will only talk about how you can cope with the situation. The most effective way to protect your corn from wild animals is a good, smart dog patrolling your plantings. A second way is with electric fencing, a single wire a foot off the ground for animals the size of raccoons and another about 4 feet high for deer. I do not guarantee that one wire 4 feet high will stop deer, but that's what some gardeners maintain. In this situation, the deer are mincing daintily around the garden and rub against the fence or examine it with their noses. Wham. They don't come back for a while. Some say that if the wire is baited with aluminum strips smeared with peanut butter it will be even more effective. The deer, presumably, lick the peanut butter, get a jolt, and never come back. But don't bet on it. The weakness of electrified fence is its fragility. When deer are running through a big field, as they often do, they are apt to run right through a single strand of wire and drag it back to the woods. Or jump it if they see it. Deer can jump about 8 feet over a fence if they are hungry enough.

On small garden plots, I now set 8-foot posts around the perimeter and hang lengths of woven wire fence, cut from old, discarded fencing, on the posts. Around the bottom, I stretch chicken wire 2 feet high to keep out rabbits. The woven wire hangs above the chicken wire. I keep live traps set all the time, to catch raccoons. Then I kill them. If I do not kill them, about every few years distemper will infect dense coon populations and kill them. If I tried to wait for the distemper to do the job, the food-production system on our farm would be ruined. For reasons I can't explain, the fur trade is paying good prices again (fall 2008) for raccoon pelts, as much as $25 each, in winter, in season. Perhaps this will motivate coon hunters to keep raccoon numbers from exploding, as hunters did for many years in the past.

With the rabbit wire at the bottom, and the woven wire above, the fence is about 7 feet tall and so far the deer haven't jumped it. Near Hudson, Ohio, I know a farm that has gone to the expense of putting up a deer-proof woven-wire fence, with electrified strands at the top, 8 feet off the ground around twenty or more acres. This is very expensive, but deer are so bad there (a hundred per square mile, the wildlife specialists tell me) that farming would not be possible otherwise. This farm can only afford it because it is selling highly priced fruits and vegetables directly to consumers.

Although I much prefer using "humane" ways to keep animals from destroying food, the logic behind it is shortsighted. So you keep deer off your farm. They go to some other farm. If they can't get to any farm, or garden, or ornamental planting, they will eventually start starving to death, as they are in many places.

Some people will swear by various concoctions sprayed on plants, or hung over the corn plants that are supposed to repel deer and raccoons with their odor. I think I have tried them all, including human hair, various kinds of evil smelling soaps, mothballs, tankage bars, human urine (but not tiger manure from the zoo as some swear by), and all I can tell you is that a hungry deer or raccoon will not be dissuaded for long. Some of the new ones actually work but require frequent reapplications. My wife tried the ultimate in scare tactics last summer. She parked the car out by the sweet corn. She put the keys on the night table in our bedroom. Several times in the night, she rolled over sleepily and pressed the panic button on the car keys. All hell broke loose in the garden. Yes, that worked. It also set every dog in the neighborhood to howling, and inevitably, their owners did a little howling too.

Deer are developing a taste for corn silks. In our county, especially where cornfields are adjacent to woodland, the deer bite off the silks before the ear can be pollinated. They come back later to bite off ears where they did not bite off the silks earlier. Raccoons take their toll too and seem to know exactly when the corn is perfect for roasting. Wild turkeys like to eat the newly sprouted corn, then come back at harvest time, knock down the stalks with their breasts and wings, and gobble on the ears. They may take half an acre from a field corner. Gobble gobble gobble. On a two-thousand-acre corn farm, the damage from these three corn lovers does not seem too alarming yet, but every year there are more gobblers. Herds of deer in our county sometimes number thirty and forty animals each (I've counted them) wandering at will over the farmland. Only when these herds finally stampede down city streets will our wildlife-loving society realize something has to be done.

Birds of various species, especially crows, like to peck sprouting corn out of the ground. Fortunately this kind of predation is not usually severe. I used to coat seed with tar when planting small patches, although that was time-consuming and not very effective. I have resorted to scarecrows, which actually work a little. My scarecrows are simple affairs. I stick a tall stake in the ground, then hang a horizontal stick by a piece of twine on the vertical stake. I put an old shirt on the horizontal stick like you would put it on a clothes hanger. Then I put an old straw hat on top of the stake and tie two pieces of aluminum foil to the ends of the two shirtsleeves. Another piece of aluminum foil goes like a face under the hat. When the wind blows, the aluminum foil in the shirtsleeves swings about. The aluminum foil under the hat looks sort of evil, a paleface. Because this scarecrow is sort of mobile, it actually scares birds, even deer for a while. By the time they catch on, the corn has grown a little beyond sprout stage and the birds don't bother it anymore.

Flocks of red-winged blackbirds can be very troublesome in ripening corn, but most other birds eat only the ends out of a few ears, and pay for their meals by eating insects that attack corn. Squirrels, opossums, and skunks also like to nibble on sweet corn as it ripens. We have lots of old socks which I slip over the ears. Doesn't take long in a garden patch and it helps.

Where I live here in northern Ohio, one of the worst insect pests of corn is comparatively new here. The Japanese beetle has finally made it this far from the east coast and, as is generally true of new pests, it has reached very damaging population levels. These high levels will decline, if experience holds true, but right now these insects are attacking corn silks and eating them before the ear gets pollinated. Commercial farmers are spraying insecticides to stop them. So far, I just cross my fingers and hope they don't get too many silks or don't eat the silks until after the ears are pollinated. The common wisdom says that beetle traps aren't effective, but we think they help. In our garden, my resolute wife also patrols the corn in the evening with a can partially filled with gasoline and knocks the beetles into it.

The European corn borer poses, or at least posed at one time, the worst threat to corn in the insect world. When it first struck—savagely—many years ago, the agricultural world panicked. Dire predictions of the end of the corn industry were made wherever podiums and after-dinner speakers got together. If they had been available, chemicals of great toxicity would have been hurled into the fray, but fortunately (in the long run) few were available.

Instead corn growers learned how to control the borer with resistant varieties and good cultural practices. But the borers are still out there, and no variety is completely resistant to them. Some new genetically modified varieties are actually poisonous to borers. Sounds great, but do you really want to eat something that kills borers when they eat it? Fall disking or plowing and stalk shredding all help control the pest since it overwinters in stalks. Natural predators and its own diseases have contributed to corn-borer control too. Some farmers still spray insecticides to control it, especially the second generation that attacks the silks and hinders pollination, but for the most part, the practice does not pay in field corn. Damage is not that bad anymore. (As my luck goes, as soon as that statement goes into print, a new outbreak will occur.)

In sweet corn, commercial growers will often spray for earworms because the consumer—the ones who love wildlife mammals but not wildlife worms—won't tolerate an occasional bug in the corn she buys. She'd rather have corn with poisons on the husk than a harmless worm. The organic grower on small plantings can alleviate the worm problem a little by putting a drop of mineral oil on the silk of each ear. On large plantings, the only organic control of earworm damage on the ears is to alternate plantings so that your main crop matures between the first and second worm generations. You'll have to experiment to find out the proper timing in your own locale and keep notes. Eventually, you can determine which of three plantings of corn is freest of damage. Make that your main crop, if you intend to sell corn. Earworms are more of a problem in the southern and central states. Where temperatures fall below zero, most overwintering earworms die. New infestations come in the spring from more southern climes but usually don't arrive here in northern Ohio until late in the season.

Corn rootworms (southern rootworm, northern rootworm, and western rootworm—all distinct) are a serious pest of corn where commercial growers insist on planting corn on the same land, year after year. Now that the cheaper chemicals used to control rootworm and other soil insects in continuous corn are banned there may be a return to the sanity of crop rotation by all corn farmers. All the rootworms feed on the roots, as their name indicates, causing wilting, weakening, and even death of the plant. The southern corn rootworm is the least dangerous, but as an adult beetle, it becomes the striped cucumber beetle, a painful thorn in the side of organic melon, pumpkin, squash, and cucumber growers.

Wireworms like to dine on young corn roots too, and the seed itself. Rotations won't control them either, but fortunately, they rarely get to epidemic proportions. Growers have discovered that wireworms rarely move out of the row they attack, which is kind of a mystery. You can replant a new row down between two rows that have been attacked by wireworms, and the new corn will remain unscathed. The timing probably has something to do with this phenomenon. Cultivation controls wireworms somewhat. They thrive better in poorly drained soils and cool, wet springs or in no-till fields.

Seed-corn beetles and seed-corn maggots eat the germ out of the seed but are not serious threats. Treated seed controls them, if necessary. Treating seed with coal oil or tar can discourage corn beetles and maggots, but if you use organic methods you would need to check with your certifier to see if this practice is legit. Since the coal oil/tar regime isn't practical on larger plantings, and only partly effective in any case, it's best to avoid it.

Armyworms can harm corn sometimes, but not even chemical farmers believe spraying is usually necessary. The spray kills too many predators of the armyworm. Corn-leaf aphids may appear in great numbers some years, but again, even in the opinion of chemical farmers, spraying is questionable. Corn aphids come and go, don't seem to hurt corn yields, and spraying them kills too many ladybugs and other predators. If you should see thousands of small black bugs swarming up the stalks of corn around the outside of your field, that's the chinch bug. Infestation is seldom severe, though the bugs will weaken or kill plants they feed on in great numbers.

Corn Diseases

I feel obliged to list some corn diseases here, but be of good cheer. You can go through a gardening life and never be affected much by any of them. Corn, as I say, is tough stuff. The seven deadliest sins in this category are southern corn leaf blight, which scared the wits out of corn producers in 1970, northern corn leaf blight, northern corn leaf spot, maize dwarf mosaic, diplodia rot, gibberella rot, and bacterial wilt. A picture of a serious infection of any of these diseases is enough to make you ill. And in fact, some of the molds that rot corn *can* make you ill. Don't ever eat moldy corn. Or feed it to livestock. In addition to poisoning that might result, especially from gibberella, there's always that chance in a thousand that moldy corn, especially moldy white corn in the South, might be infected with aflatoxins known to be carcinogenic.

Fortunately, plant breeders have been able to outmaneuver these seven diseases with at least partially resistant varieties. Sometimes, though, the outmaneuvering backfires. For instance, the devastating effect of southern corn leaf blight in 1970 can be blamed at least partially on the corn industry's own carelessness. Because it made very desirable single-cross corn varieties, Texas male-sterile cytoplasm was being used to produce almost *all* hybrid seed corn, even though this meant that any disease to which this kind of cytoplasm would be susceptible would have the potential for wiping out most of the crop. The fungus that causes southern corn leaf blight was well known and thought not to be a threat. What was not so readily recognized was an earlier report from the Philippines that showed the disease seemed to be particularly virulent there against corn with Texas male-sterile cytoplasm. Scientists assumed it was only weather conditions in the Philippine Islands that were responsible.

As a matter of fact, what had erupted there was a hitherto unknown race or strain of the disease, now known as Race T. When Race T inevitably reached the United States and our vast fields of corn, humid weather sent it racing through the farmlands like a forest fire. Plant breeders found themselves almost literally in a race with potential starvation to breed enough seed corn without Texas male-sterile cytoplasm for the 1971 planting season. About the only people who could breathe easily at that time were the handful of growers who raised non-hybrid corn. That's when I started.

Hybrids are being developed with resistance to both southern and northern leaf blight. The newly discovered northern corn leaf spot is related to the other two blights, but attacks both normal and male-sterile cytoplasm. Fortunately, it has not caused serious damage, and plant scientists hope that plowing under plant debris, rotations, and other clean-culture practices will control the disease until resistant inbred lines can be developed.

Maize dwarf mosaic (MDM) is a serious corn disease wherever corn is grown in the proximity of Johnson grass, which is one of the worst weeds of the mid-South and a host for the disease. More recently, another strain of MDM has been found in the East that doesn't seem to be connected with Johnson grass. Aphids transfer the virus from weeds to corn. It can even be carried in clothing or on farm machinery from field to field. The leaves of infected corn become finely stippled and the plant is stunted. Sometimes there's a proliferation of shoots from the base of the plant too. Your only defense is to plant genetically resistant varieties and get rid of Johnson grass, if you can.

Diplodia rot strikes stalks and ears, especially in early corn varieties. If nitrogen levels are too high in relation to potassium, stalk rot is more likely to occur. Then, even if the ears are good, the stalks fall over, making harvest difficult if not impossible. When diplodia attacks the ears, they become covered with a whitish mold and do not mature properly. A very dry early season followed by wet weather later on increases the chances for a diplodia infestation, but seldom is the disease severe.

Gibberella rot, like diplodia, attacks a plant as it matures, not when it is young and growing. Resistant varieties and clean plowing are standard controls. Gibberella can be distinguished from diplodia because the former infection almost always has a reddish color to it, both in the stalk and on the ear. Gibberella-infected corn should not be fed to swine, but the experts say it "may be" fed to chickens or cows. Not my chickens or cows.

Bacterial wilt or Stewart's wilt (or Stewart's disease) you are more apt to find in your sweet corn, especially in the Northeast. Yellow lesions appear in late spring on young corn plants, running parallel to the leaf veins along the length of the leaves. The infection looks like a number of other corn diseases, or nutritional deficiencies, so you'll probably need an expert to identify it. But if your ground is reasonably fertile and well drained and the season is normal, long yellow lesions on the corn leaves strongly suggest this wilt. The disease is caused by bacteria that survive over winter in the bodies and mouthparts of the hibernating corn flea beetle. Corn flea beetles are shiny black, about 1/10 inch long, and hop like fleas. Severe winters will kill hibernating beetles and therefore reduce the bacteria too. Otherwise, the best defense is to plant resistant hybrids. There are many good wilt-resistant sweet corn varieties for the East, especially among newer varieties.

Tips on Harvesting Corn

About seventy-five days after a sweet corn variety that is supposed to mature in seventy-five days is planted, the corn should be ready to pick. Notice all those conditionals. The only way to know when the corn is actually ready is to pull down a sliver of husk and take a peek. Jam your thumbnail against a kernel. If it pops and the milk squirts out, the corn is just right. If your fingernail goes through the skin of the kernel easily, the corn is a little green yet; if you must press quite hard to penetrate the kernel, it's too old. After you've picked corn awhile, you'll be able to tell by the condition of the silk and the fullness of the ear. But like everything in farming, that advice should be followed by the phrase, "It depends." I like my sweet corn a little on the pimply side, a little green yet. My biggest gripe with commercial sweet corn you see in produce markets is that it has been picked a little too late for my taste and is a little tough and gummy. When you bite into an ear and the juice squirts across the table and hits your companion, it is just right to eat, in my opinion.

To harvest small plots of popcorn, it's best to jerk the ears from the stalk, leaving the husks attached. Peel the husks back and use them to tie five or six ears together. The ears can then be draped over a wire by the husks to hang and dry in the barn or attic. To keep mice from getting to the corn, poke a hole in tops of tin cans with a nail, then slide the tops on the wire, one on each end about a foot from the wall. The mouse may be able to walk a wire, but it can't pass the tin lid blocking the way.

Harvesting ears of corn is most easily accomplished with the help of a husking peg. This device comes in a variety of styles, but its purpose is to rip open the husks quickly, with a minimum of wear on the fingers.

The easiest way to harvest field corn by hand is to allow it to hang on the stalk until the plant is dead and brown. Then move down one row after another, husking the ears and tossing them onto a wagon, cart, or pickup truck. Breaking the ear of corn out of the husk is a skill you will develop as you go along. The way I do it is to grab the ear at the base with my left (gloved) hand, while my right hand strips away the husk and bends the ear back between the thumb and forefinger of my left hand, snapping the ear free at the stem. The drier and riper the corn, the easier the ear will snap out of its husk. I use a husking peg in my right hand to strip down the husk.

In the past, farmers used many kinds of husking "pegs" strapped to their fingers or the palm of one hand, the purpose of which was to rip the husk loose from the ear with one swift downward motion. The original husking peg was a simple piece of wood or bone about a ¼ inch in thickness, sharpened to a point at one end, and held in the middle joint of the fingers of the right hand. A string or leather thong held it to the fingers. Grasping the peg the way you would grasp a knife if you were slicing potatoes, the

Harvesting corn by hand.

husker would slash open the husk with the point of the peg, grab a section of husk between thumb and peg, and tear it off the ear. You can do the same with your fingers, but it takes longer and is hard on the fingers.

Improvements came to the husking peg, the ultimate being a sort of leather, fingerless glove with a steel plate riveted to the palm part of it. The plate had a steel hook on it, which ripped open the husk when the husker ran the palm of his hand down the ear of corn. A good husker could perform this operation with amazing speed and skill, and could, in fact, husk a bushel of corn from the stalks in a couple of minutes.

Husking pegs are still being made and used. Check at rural hardware stores. Corn-husking contests are coming back into vogue in rural areas, usually held at county fairgrounds in the fall. You will find all kinds of husking pegs on display at these events. But the very best bet is to check out sources on Google or the search engine of your choice. There is really no need to put sources of equipment into a book anymore as there is nothing, absolutely nothing, that you can't find on the Internet.

After you husk the ears from the standing stalks, you can let sheep, cows, or horses eat the leaves (fodder) off the stalks, then break up the stalks with disk or stalk shredder or rotary mower for incorporating into the soil as organic matter. Or you can cut the stalks when the leaves are still a little greenish and the ears still in place on the stalk, tie the stalks into bundles (about a dozen stalks to a bundle), and stand the bundles up into shocks. This is the cheapest and best way to dry and store your corn right out in the field. Shocks of corn are what you see in pretty calendar pictures but rarely anyplace else these days except on Amish farms.

For cutting corn by hand you need a corn knife. Corn knives look just a little bit like swords, and old ones sell at almost every farm sale as antiques. But you can buy them brand-new too. Every rural hardware store should be able to order them if they don't have them in stock. Or consult the Internet.

You cut the cornstalk off about 4 inches above the ground with a short downward stroke, after having grabbed the stalk up high with your left hand. (If you're left-handed, reverse hands.) Continue down the row, gathering each stalk in your left arm, and cutting it off with the knife in your right hand, one-two, one-two. A definite

A corn harvesting crew in front of a long row of corn shocks on the farm, with a new crop of fall-sown wheat growing in the foreground.

rhythm will assert itself as you swing along. Just be careful not to get overwhelmed by the spirit of the thing and hit your leg with the knife. I can't resist telling a funny story. I once decided to cut the lily pads out of our farm pond by wading into the water and slicing them off with a corn knife. All well and good in the shallower water. But all at once I stepped off into the deepest part. Have you ever tried to swim with a corn knife in one hand?

When your arm is full of stalks, drop them in a neat bundle on the ground. Later you will tie the bundles with twine and set them up into a shock. You don't really have to tie all the stalks into bundles. Instead, tie maybe six bundles for the nucleus of each shock. Once the shock was started with these bundles, set armfuls of cut stalks not tied into bundles against the shock. When the shock is as large as you want it, tie the whole together with twine. This method saves much time over tying all the bundles.

Setting up a shock properly requires practice. You can lean two bundles against each other, then set up two more against the first two but on opposite sides, so that all four exert more or less equal pressure on each other from all four directions without falling down. All should slant outward a little, but not too much, from

Tied into bundles, the corn stalks are shocked in the field until used—the ears and foliage as livestock feed, the stalks as bedding.

bottom to top. Think tipi. Once you get the first four bundles standing, it is fairly easy to lean the rest around that central core. After you have stood up about eight bundles, the shock will be fairly steady and you can stack more bundles against it firmly without fear of the whole thing falling over. But until you catch on, your shocks will fall over in the early stages of building, and you will be tempted to use language that could dry corn faster than artificial heat.

So here's what the old-timers did, so as not to singe the corn. Go into the standing corn where you want to put up a shock and, before you cut any stalks, bend over four stalks from two adjacent rows equidistant apart and tie them together at the top by wrapping a length of supple, green stalk around those four stalks. Bend the tie stalk over itself so it can't flop loose. Now you have the beginnings of your tipi, uncut stalks that won't fall over as you stand bundles against it.

The size of a shock is up to you; twelve to sixteen bundles makes a solid-standing structure. Tie the whole thing together a little below the tassels with a piece of twine so that rain runs mostly off the shock, not down into it. First I throw a rope around the shock, pulling it as tight as I can; then I tie the twine around and release the rope.

Once you tie the whole shock together, it will stand all winter and shed water if you have placed the bundles evenly around the shock. A handy way to make sure that air can get inside the shock to help dry the corn and fodder is to show your kids how to burrow inside the shock, make a roomy center inside, and pretend that they are Indians in a tipi. As children, we used to spend hours doing that.

The shock standing in the field becomes a "corn crib" in reality. Inside, the ears are still on the stalks, and dry out nicely. Don't be discouraged if your shock starts leaning over in the subsequent weeks. There is art in all things agricultural and it takes time to learn how to make a shock that stands erect into winter. Even Amish shocks often slant over during the winter.

After the rush of fall harvest work is past, take the shocks apart—one bundle at a time—and husk the corn out of it. The corn can be cribbed or fed, the fodder fed to sheep, cows, or horses. Even hogs. The leaves inside the shock will still be somewhat green, though very dry, and are relished by livestock. The tougher lower stalks that animals often won't eat make passably good bedding. Bundles of stalks make good insulation to keep farm buildings, especially chicken coops, warmer in winter. Simply lean the bundles up around the building.

After harvest—of whatever kind of corn you grow—you will have the stalks in the field to contend with unless you have shocked the corn. Even then, you will have the heavy root system of the corn plants still in the ground. Heavy plows or disks will work these stalks and roots up nicely. Smaller plows and disks, the kind you will most likely be using if you've planted only an acre or two of corn, will handle the roots, but not the stalks too well. If possible, shred the stalks with a stalk shredder or any rotary mower. Do the same on garden plots before you rotary-till the ground for next year's crop. Shredding not only makes tillage easier and increases rate of decomposition, but also helps control any insect pests harbored in the stalks. Heavier rotary tillers for farm tractors, like the Howard Rotovator, will chop and incorporate stalks into the soil too. You can plow under, disk in, or otherwise incorporate stalks in the fall or spring when the soil is fit to work. If you have sown a cover crop on the field, you won't plow, of course, until spring. Farmers often used to sow fall wheat after corn, in which case they plowed or disked the ground immediately after the corn was off, and then planted wheat in October. Instead of bare ground all winter, the field would be carpeted with wheat.

Many farmers do not believe plowing is the best method of cultivation. They claim the offset disk incorporates organic matter throughout the topsoil whereas the plow buries plant residues too far below the soil surface. Besides, plowed ground erodes worse than disked ground, and few soil experts will disagree with that. (The argument over the pros and cons of plowing is at least a century old.) I'm not sure the criticism of the plow should concern garden farmers because they will be using the plow (if they do) mostly in late spring to turn under green manure crops. Late spring plowing followed by planting does not cause as much erosion as fall plowing does. As for burying the plant residues too deeply, *small* plows, one or two bottom plows of lighter weight, won't do it to the degree modern large plows will. In the garden, the dispute is of less importance since the rotary tiller will be used most of the time instead of a plow.

Storing and Using Your Corn

Weevils will attack corn, or so I've been given to understand, but I've never observed them in ears of unshelled corn stored for just a year in traditional, slatted cribs. If weevils do harm corn, the problem is far less critical than it is with cereal grains, another reason why corn is the best homestead grain for you to grow. Because of the way corn is stored you will, however, have to contend with some rodent damage. But rats in grain bins are easier to control than bugs.

Like everything else in agriculture, there's no one right way to store corn. You can approach "rightness" in agriculture in direct relationship to the amount of time or money you want to spend. Our society believes that if it's "right" to drive a Ford, it's "righter" to drive a BMW, but it ain't necessarily so.

The most primitive and cheapest method to store corn is to cut and tie the stalks, ears attached, into bundles, set the bundles into shocks, and leave them in the field until needed, as previously described. The shock becomes the storage building, so to speak, effectively keeping rain off the unhusked ears inside while the grain dries down in a month or more to proper moisture content (about 13 percent) for long-term keeping.

The bundles of stalks, ears still attached, can be fed to livestock. They'll eat the grain and the dried leaves (fodder) off the stalks. The ears can be husked from the bundles and fed, shelled or unshelled, to hogs, chickens, horses, and cows, even rabbits. The bundles, divested of their grain, make good fodder feed for sheep, goats, and rabbits.

The disadvantage of storing corn in the shock in the field is that you can't haul it out when the ground is muddy in winter. And when the ground is frozen, the butts of the shocks freeze in the mud and are devilishly hard to kick loose. In great-grandfather's time, farmers found at least one way to circumvent the problem: They hauled the bundles to the barn in late fall, and husked the corn at their leisure during the long winter evenings. You can too, especially with a small planting.

That practice led to what surely was one of the most pleasant customs on yesterday's farms: husking bees. Young and old would gather together, first at one farm, then another, and husk the corn. Actually, the husking was merely a by-product of a social evening. The husker who found a red ear in his bundle was allowed a kiss from boyfriend or girlfriend, and of course, what more motivation does one need to keep husking? And in those days, corn did not have the dull sameness of today's hybrids, so there were quite a few ears that would turn up red. And quite a few more that were tinged with enough red to qualify for a kiss.

I keep asking myself if the "old days" weren't really better after all. There were no sharply drawn distinctions between work and play, between manual and intellectual labor, between business and social life. There were more chances for people of different ages to mingle and understand each other. In this one instance—and one can name many other examples—the work of food getting was turned into fun; neither love nor labor was lost. We could do far worse today.

When the corn was shucked at the old husking bees, the husks were not always broken off the ears of corn. Often they were braided together in most artistic ways. As much as a bushel of corn could be thus tied together and hung up like clothes on a line, where it could dry properly without rodents getting to it.

There's no reason why the small homesteader could not do this today. I'm romantic enough to believe that the time spent would be far more rewarding than watching television, especially if there were a few good friends sitting there in the light of the open barn door, talking good talk. Or, even more especially, if a pretty girl were showing me how to braid the corn husks. I have a notion we have lost some of our ability to enjoy life to its fullest today, but that notion is supposed to be a sign of old age. I'll leave you with a thought: One more reason for growing open-pollinated corn is that you never know when a red ear and someone you love might show up at the same time.

You might braid and hang up your popcorn at least—and sweet corn for parching—away from rodents, but most garden farmers will require a crib to store their corn for livestock. There are several ways to build one, again depending on whether your desires run to Model T Fords or Humvees. My first crib, long gone, I built out of saplings cut from the woods, and some junk lumber and roofing panels. I needed only a small crib, one that would store about 75 bushels. A cubic foot equals about ⅘ bushel, so a crib 6 feet by 4 feet by 6 feet tall was adequate. Using saplings 3 to 4 inches in diameter, I tried to duplicate roughly the requirement of the old traditional, slat-walled crib to allow for good air circulation to dry the corn. I slanted the walls outward a little as they rose so rainwater would drip down and away from the crib. (Rain wets a little of the corn through the slats, but not enough to harm it as long as the grain can dry out quickly after the rain stops.)

For a floor, I recycled a 4 × 8-foot panel of ⅝-inch plywood that had been previously used for a sign. Eight old concrete blocks scrounged from a junk pile became the foundation under the floor, one block at each corner and one on each side midway between the corners.

First I set four white-oak posts into the ground at what would be the corners of the future crib. Catalpa or tamarack or locust would have been better, but white oak lasts about as long as I figured the crib would. The posts were about 5 inches in diameter, and I sank them about 2 feet into the ground, leaving some 7 feet of length above ground. The space between the posts measured a little less than 4 feet widthwise, and a bit over 6 feet apart lengthwise.

Across the top of the corner posts, both widthwise and lengthwise, I nailed 2 × 4s to hold the posts rigid and to form support for the roof. I made sure that the corner posts all leaned outward a little so the subsequent walls would also lean a little outward to shed water, as I explained. That's all the nailing I had to do except for the tin roofing I applied later.

Then I started building up the walls from the floor in a manner similar to laying up logs for a log cabin. The only difference was that my sapling logs had to be cut continually longer as I progressed upward so that they lapped the slanting corner posts. To make sure the saplings would be long enough to lock in place against the corner posts all the way up, I cut them extra long. Then, when the walls were finished, I sawed off the excess lengths with a chain saw. When filled with corn, the log walls became tight and immovable; the pressure of the corn inside the crib held the sapling ends firmly against the uprights as securely than any nailed wall could do.

The space between the logs ideally should have been about 2 inches, but even in the straightest saplings, there were crooks and curves that occasionally left a space wide enough for an ear of corn to slip through. But no matter. When you fill a crib like this one, the ears of corn will flow against the walls and eventually block the wide gaps. You have to pick up a few ears that fall to the ground, but not many.

However, sometimes you may have to notch a sapling end into another as you would in a log cabin, to narrow the gap between the saplings sufficiently so the corn won't fall through too easily. Also, be sure to alternate a butt end and a top end at the corners as you lay the wall up to keep it fairly even and level with the world.

At the top of the crib at one end, I left an opening for filling. I hauled corn in from the field in my pickup and stood on the pickup bed to shovel the corn into the crib. At that height I could easily toss the corn into the opening as I do now with a regular, slat-walled crib. To take corn out of the crib, I removed the saplings from the top down as the crib emptied.

I built the crib in one afternoon. Its only fault was that it was not animal- or bird-proof. Squirrels, chipmunks, jays, and mourning doves feasted all year long. Coons were the worst. Fortunately there was enough corn there for both my wild and domestic animals, but later on I built the crib I use now, which keeps out varmints most of the time.

Another low-cost crib you can build is with snow fence or picket fence. It doesn't require a whole lot of skill either. First you build a solid floor about 6 inches off the ground, boards over planks or posts or old telephone poles laid right on the ground in a level spot. The floor should be big enough to accommodate a circle

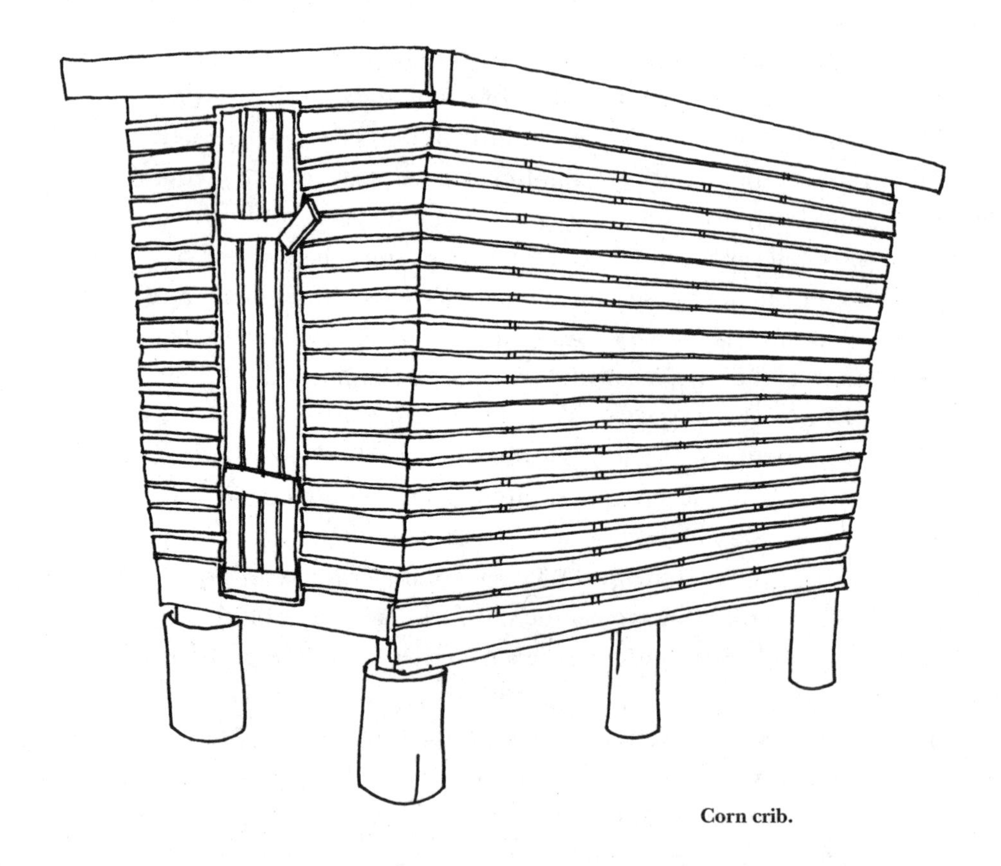

Corn crib.

about 15 feet in diameter. Then make a circle of snow fence or picket fence approximately 14 feet in diameter so it will fit nicely on your floor. You'll need 154 feet of fence to complete the circle. In the very center of the enclosure set another piece of fencing measuring a foot or so in diameter. This small middle piece becomes a ventilating shaft allowing air to penetrate to the middle of the corn you will be piling around it.

Once the enclosure is full of corn, you can add another circle of fencing on top of the corn, a foot or two smaller in diameter than the first. Set another small piece in the center to extend the ventilation shaft up through what will be your second layer of corn. I've seen third layers stacked onto some of these makeshift cribs, but that's risky. A crib can get top-heavy if constructed this way, and topple over.

Whether you build a one-, two-, or three-tiered snow-fence crib, you will want to cover the corn. Bundles of cornstalks laid closely together over the corn, tops all together in the center and raised slightly so water will run down to the butt ends and off onto the ground, not the corn, will give the grain adequate protection. I've seen pieces of roofing tin laid on top and weighted down with posts. You don't need anything elaborate, and after you have had a chance to experiment on your own you will probably come up with other makeshift materials.

Getting the corn out of a snow-fence crib isn't easy. You usually have to knock a hole in the fence at floor level big enough to accommodate a scoop shovel. Standing outside the crib, you shovel as much of the corn as will spill down and out your hole. Sometimes the corn will bridge over above your hole and you have to ram a crowbar through the fence into the corn to jar it loose. When the crib is emptied to the point where you can't shovel from the outside anymore, you crawl inside and pitch the corn over to your outlet hole.

Farmers usually use small, motor-driven grain conveyors to make unloading the cribs easier. The conveyor hopper is placed right outside the exit hole of the cribs. Corn is shoveled into the hopper and elevated to a waiting truck or wagon.

Yet another low-cost, but more permanent type of crib can be built out of telephone poles or large posts. Quite large cribs can be built with the former, but you'll need mechanical help, at least a tractor forklift, to lift the heavy poles into place. For most homesteads, 15-foot creosoted posts from the lumberyard will make a crib sufficiently large for your purposes.

The pole crib can be built as long as you wish to make it. The limiting factor is width; don't make it wider than 4 feet, so as to allow for good air penetration. Simply set the posts solidly in the ground, a pair of them 4 feet apart every 6 feet. Nail boards (1 × 6-inch boards are okay, but 2 × 4s are better) around the top of the posts for extra strength to keep the posts from sagging.

Nail 2 × 4-inch crosspieces between every pair of posts about a foot above ground level, then nail boards to the crosspieces to form the floor of your crib.

Then staple a strong, close-meshed wire fencing to the insides of the posts to form an enclosure that is rat- and bird-proof. Second choice: nail 1 × 2-inch slats to the poles on the inside, each slat about 2 inches apart. But of course rats and squirrels will gnaw through unless you line the wall with some kind of hardware cloth. Top off with a simple slant roof of corrugated steel or aluminum.

A door should be framed into one or both ends of the crib for filling and emptying.

If you want a crib that preserves the corn better, you can build the type traditionally found on American farms, as I did. Thousands of them are still standing, though often empty. You can get a good idea of how to build one by taking a ride through the countryside and observing them from the road.

Common to all of these cribs is the 4-foot width and the spaced slats of the walls for ventilation. The cribs may have a simple slant roof or common double-truss roof. Sometimes the walls are perpendicular to the ground; other models may angle slightly outward from the ground to the roofline. The purpose of angled walls, as I mentioned earlier, is to allow water striking the wall to fall down and away from the crib, minimizing the amount of corn that gets wet from the open slats.

There's a proper way to make a door in a corn crib. You can get by with an ordinary hinged door, but if you intend to fill the crib full you have to have that door closed; that is, you will have to have another opening through which to fill the crib. Moreover, once the crib is full, you won't be able to open a regular door because if you do, the corn will come rolling out.

So farmers long ago learned how to solve the problem. On the sides of the door frame, they nailed small pieces of wood, angled downward from outside to inside of the crib. These wooden strips form slots into which 1 × 4-inch boards the width of the door opening can be slipped. When all the boards are in place, closing in the door opening, they look like a set of shelves tipped upward rather than level. The boards are close enough together to hold in the corn, but they can be easily slipped up and out of the slots. Because of their angle of repose, the weight of the corn does not pinch them tight. When you want to take corn out of the crib, you remove the top board, then keep removing boards from the top down until the crib is empty enough so that corn no longer rolls against the door. Or you can slip out the boards at the bottom to let the corn roll out or your shovel to scoop in.

You can also buy steel corn cribs. Used ones, now obsolete in commercial farming, dot the countryside, and if you can hire the machinery to move one of them you can often purchase them for a reasonable price.

Whatever kind of crib you choose, some rodent damage is inevitable, though in the steel cribs and the well-constructed pole or traditional cribs you can keep it at a minimum. But it pays to keep a cat or two around the corn crib. I built my slatted board crib up off the ground on pillars about 2 feet high and encased the pillars in old tin roofing. The rats can't climb up the pillars.

Corn Silage

Corn silage is green corn chopped, stalks and all, when the kernels are just beginning to dent or harden. The chopped material is packed tight in a pile so that oxygen can't get to it freely enough to decay it. Storage buildings can be the upright silos you see on most livestock farms, or "trench silos," or nothing more than huge plastic bags of the chopped ensilage closed tight to keep out oxygen. On a small homestead, silage is not a very practical idea unless you happen to have bought a farm with a small silo already on it, or if you live so far north that your corn in some particular year does not have time to ripen before frost.

To make silage you need some kind of ensilage chopper to chop up the corn, stalks and all. Also you need a blower if you intend to put it in an upright silo. To avoid the cost of a silo, some livestock feeders and dairymen bulldoze a trench in a hillside and fill that with silage, packing it down tight with tractors. The silage on top of the trench—about a foot down into the pile—may spoil from contact with air.

Silage is not very good livestock feed in my opinion, but you should be aware of this method of using corn. For winter feeding on a small homestead, it is cheaper to make shocks as explained above and feed the dry fodder. Sweet-corn fodder is especially palatable to cows and sheep. When I have turned sheep into stands of dry, dead sweet corn, the sheep eat even the thicker stalks close to the ground.

Shelling and Grinding Corn

You will have to shell your corn for your own table use and perhaps for your animals. The Nasco Farm & Ranch catalog (Fort Atkinson, Wisconsin, 53538 or Princeton Ave., Modesto, California, 95352) lists a hand-cranked sheller capable of doing maybe 10 bushels an hour, and a small hand sheller you might find adequate for shelling a few ears of popcorn for an evening. Popcorn is hard to shell off with your fingers. We select seed from our field corn for easy hand shelling, and I do it when feeding a couple of ears to the hens or getting enough for a cornbread dish. These shellers are available from many sources. (Again, the Internet to the rescue.) Larger, tractor-powered shellers are still available too. Some searching of used farm machinery lots might help you locate a good used one, since shellers are obsolete in commercial farming. The new corn combines shell the corn as it is harvested, so shellers are no longer needed on most farms.

Motor-powered grinders or mills for making meal for animals are also commonly available from farm-machinery dealers, or you can take corn to be ground to commercial feed mills and/or elevators. I am a little prejudiced against grinding corn for animals. They can handle whole corn about as well as ground corn. If you put some hog manure in a bucket with water, stir it up and pour it through a sieve, you will find that much of the milled yellow outer coating of the corn has gone right through the animal just like the outer coating of sweet corn goes through you. So the argument that milling makes the corn more "digestible" is not really correct. It just means the animal will eat more and make the miller more money. It seems to me that we citizens of mighty America, the most powerful nation in the world, have the attitude that to handle whatever confronts us, first it is necessary to pulverize it.

Small kitchen grain mills of all kinds are very popular now and easy to find in catalogs or on the Internet. You can grind corn into meal with your blender, but the process is hard on the blender blades and you'll probably graduate quickly to one of these mills. Hand-cranked or electric models are available. Make your meal from either yellow or white field corn. And believe me, you'll recognize that your own freshly ground product from the current year's new corn, white or yellow, tastes better than any commercial cornmeal.

Further Tips on Processing Corn for Table Use

If your popcorn doesn't pop well, it is probably too dry. If the unpopped kernels are dark and scorched, with a lot of partly split kernels, and if a muffled pop during the popping is the best noise the corn can muster, the corn's too dry. Add a tablespoon of water per quart jar, seal again, and shake well twice a day for a couple of days. Try popping again. If the corn is still too dry, repeat the treatment. When corn is too moist, it will pop with a relatively loud explosion but popped kernels may be small, jagged, and tough.

Real popcorn gourmets say you can't pop decent corn in an electric popper. They're too slow. The corn gets tough. You have to use a hand-cranked popper. Get your burner red hot. Set the popper on it and quickly pour enough oil to cover the popper bottom thinly. Household cooking oil is preferred by most people, but try olive oil or peanut oil sometime. The oil is smooth on the surface when you first pour it in. When you can see broken lines sifting across the surface of the oil, and smoke is beginning to rise, add the corn and quickly spread popcorn salt over the corn, 1 tablespoon per ½ cup of corn. Immediately start rotating the stirrer.

The corn will begin popping quite soon, and will pop very fast. The real impresario will have put enough corn in his popper to raise the lid at least 2 inches above the brim of the pan, and will dump it into the eating bowl even as the last kernels pop. Waiting for every last grain to pop over the stove reduces tenderness.

Add butter and, oh mama, what a treat! Some country connoisseurs use lard, but it has to be home-rendered new lard, not that stuff you buy in stores.

Do not forget the lowly corncob when you are thinking about eating corn. Corncob jelly is a midwestern favorite. Corncobs are also great for smoking meat, and for starting fires in a wood-burning stove. Four or five dry cobs are just as good as most of the fire starters you have to pay good money for.

Corn Recipes

A few recipes are offered here just to give you an idea of the possibilities, but I highly recommend that you consult your favorite cookbooks. There are so many dishes to be made from corn. Our favorite cookbook for grains is an old one from 1951, *Whole Grain Cookery* by Stella Standard. Although it does not even get into the many foods you can make from sweet corn—fresh, canned, or frozen—it contains sixty-three recipes just for corn and cornmeal. Also check out Mexican cookbooks for corn recipes, especially those that involve using corn husks. Mexico has many, many more centuries of experience with corn than we do. I also think you will find interesting reading the famous and sometimes irreverent cookbook *Nourishing Traditions: The Cookbook That Challenges Politically Correct Nutrition and the Diet Dictocrats,* by Sally Fallon with Mary G. Enig (Winona Lake, IN: New Trends Publishing, 2001).

Hasty Pudding

1 cup nonfat dry milk
3 cups water
½ cup cornmeal
3 tablespoons oil
½ cup molasses
1 teaspoon salt
½ teaspoon nutmeg
1 cooking apple, pared and diced

- Preheat oven to 250°F.
- Oil a 2-quart baking dish, with cover. Combine nonfat dry milk and water with a wire whisk.
- In a medium-sized saucepan, bring 1⅓ cups of the milk and water mixture to a boil; gradually add the cornmeal, stirring constantly.
- Remove saucepan from the heat and add the oil, molasses, salt, and nutmeg. Stir in the diced apple. Mix well and add the remaining milk.

- Pour the mixture into prepared baking dish and cover. Bake in a slow oven for 3¼ hours.
- Remove from oven and cool slightly before serving. Serve plain or with yogurt or whipped topping.

Yield: 6 servings

Corn Pone or Corn Crackers

3 cups cornmeal
¼ cup peanut flour (peanuts may be ground in electric blender ½ cup at a time)
1 teaspoon salt
¼ cup peanut oil
1 cup boiling water, plus 2 tablespoons or more until batter holds together
Parmesan cheese (optional)

- Preheat oven to 325°F.
- Combine dry ingredients.
- Stir in oil gradually.
- Add boiling water slowly, mixing with spoon and finally kneading dough. Keep adding water until dough holds together.
- With your fingers, flatten rounded teaspoonfuls of batter on an oiled cookie sheet. If a cracker is desired, flatten it as thin as possible. If a thicker, larger corn pone is desired, use more dough and don't press it quite as thin.
- Bake in oven for 40 minutes, or until slightly golden around the edges.
- Remove from pan while still warm. Cool on a rack and store in an airtight container.

Yield: approximately 5½ dozen crackers or 3 dozen corn pones

Note: Parmesan cheese can be sprinkled on top of crackers before baking them, if desired.

Corn Pudding

2 cups frozen corn (approximately ⅔ package)
1 cup boiling water (approximately)
2 tablespoons butter
2 tablespoons whole wheat flour
1 cup water
¼ cup nonfat dry milk
¼ cup chopped green pepper
2 egg yolks
½ teaspoon salt
¼ teaspoon paprika
2 egg whites

- Preheat oven to 350°F.
- Cook corn in boiling water until tender. Blend corn and liquid together briefly in an electric blender to mash it, but not long enough to puree it.
- In a large saucepan, melt butter and stir in flour gradually, until blended. Combine water and nonfat dry milk with a wire whisk, and add gradually, cooking over low heat until sauce is thickened.
- Add corn mixture to sauce and then add the chopped green pepper.
- In a small bowl, beat egg yolks and pour a small amount of the corn mixture over them, stirring constantly, and then return it to the corn mixture. Stir and cook over low heat for several minutes to allow egg yolks to thicken slightly. Add salt and paprika.
- Beat egg whites until stiff but not dry and fold them lightly into the corn mixture.
- Bake in oven for 30 minutes.

Yield: 4 to 6 servings

Corn Bread

4 teaspoons dry yeast
1 cup lukewarm water

1 cup cornmeal, white or yellow
½ cup oat flour
¼ cup soy flour
½ cup nonfat dry milk
¾ teaspoon salt
2 tablespoons nutritional yeast (optional)
2 tablespoons honey
3 tablespoons oil
2 eggs, beaten

- Preheat oven to 350°F.
- Soften dry yeast in lukewarm water and allow to stand for 10 minutes.
- Combine in a mixing bowl the cornmeal, oat and soy flours, nonfat dry milk, salt, and nutritional yeast, if desired.
- Combine honey, oil, and beaten eggs and add to dry ingredients, mixing well.
- Gradually add yeast mixture, blending well into the other ingredients.
- Pour batter into a well-oiled (9 × 9-inch) square pan. Place pan in warm area and allow corn bread to rise 30 to 40 minutes.
- Bake in oven for 30 to 35 minutes.

Yield: 6 to 8 servings

Polenta Cheese Squares

5 cups cold water
1 teaspoon salt
1½ cups white or yellow cornmeal
1 to 2 tablespoons oil
1 cup sharp cheddar cheese, grated
⅓ cup Parmesan cheese, grated

- In a large, heavy saucepan, bring 5 cups cold water and 1 teaspoon salt to a boil. Add cornmeal very slowly, stirring constantly with wire whisk or long wooden spoon until mixture is thick and free from lumps.

- Transfer the cornmeal mixture to the top of a double boiler. Place over boiling water and cook, covered, for 30 minutes, stirring occasionally. The cornmeal is finished cooking when it pulls away from the sides of the pan.
- Remove from the heat and turn the cornmeal mixture into a lightly oiled 9 × 9 baking pan with 2-inch sides; refrigerate until the polenta is stiff enough to cut (3 to 4 hours or overnight).
- Preheat oven to 400°F.
- Cut polenta into 16 squares. Arrange in an oiled baking dish. Sprinkle with cheddar and Parmesan cheese*, place in the preheated oven, and bake for 15 minutes or until cheese is melted and nicely browned. Serve immediately.

Yield: 6 to 8 servings

Note: Serve with beef stew or Italian meatballs.

** Polenta may also be prepared by adding cheddar cheese to the cornmeal mixture just before removing it from the heat. Proceed as above. Sprinkle with Parmesan cheese before placing in oven.*

CHAPTER THREE

Wheat

The Main Source of the Staff of Life

Wheat makes bread, otherwise known as "the staff of life," which alone ought to make it as desirable for both the small farmer and gardener as corn. Wheat is also an attractive plant that can add beauty and satisfaction to your gardening and homesteading efforts in addition to fiber and nutrition to your diet. Planted in the fall, it stays green until early winter and begins growing again early in the spring. In fact, with a patch of wheat in your garden, your gardening season fills the whole year, even winter, since in late February, when the snow is gone, you should be broadcasting clover seed into your wheat to become the green-manure crop or herbal remedy (red clover and alfalfa were honored for centuries as medicinal plants) after the wheat is harvested that summer. There are few landscapes more beautiful than wheat fields in November against a backdrop of brown-leaved woodland. Our county here in Ohio, like hundreds of others, resembles in late fall an almost unending series of golf courses, with new wheat shimmering like huge emeralds in the slanting sun. You can achieve the same effect on a small scale in the garden where marigolds and mums, defying frost, can form a golden border to the rich green of November wheat.

Early in spring, the wheat, which has turned brown and dormant over winter, stages its green-carpet show all over again. The plants then "stool" and send their stalks 3 to 4 feet high in rapid response to warm weather and days of longer sunlight. Within two months after the plants begin spring growth, they head out, and those acres and acres of heads bowing in the wind turn fields into rippling green seas. As the heads ripen, the plants turn yellow, then golden brown. I just don't know where you can find a picture prettier than waves of golden, windswept wheat on a side hill in June. Van Gogh thought so too. Wildlife love the cover of the standing wheat (and the eating of it too): rabbit, quail, pheasant, partridge, raccoon,

Wheat

muskrat, deer, and groundhog. And why is it that the fireflies seem to flicker heaviest in June over the wheat fields?

Kinds of Wheat

There are five commercially important wheats grown in the United States: hard red winter, hard red spring, soft red winter, white, and durum. The hard red wheats are grown mostly west of the Mississippi and are used commercially for making bread. Soft red winter wheat grows mostly east of the Mississippi and is used principally for pastry flour, but I can tell you it makes good bread too. White wheat you'll find chiefly in the Pacific Northwest (sometimes in New York and New England too), and it is mainly used for bread. Durum, grown almost exclusively in North Dakota and surrounding states, makes the flour for macaroni, spaghetti, and similar foods.

Spring wheat is planted in the spring for late summer harvest. It is grown where winters are too severe for winter wheat. The latter is planted in the fall, grows awhile, lapses into dormancy, renews growth as I've described, and is harvested in midsummer. Spring wheat does not yield as well as winter wheat, all things being equal, nor does it grade as high.

You should plant whatever kind of wheat that grows successfully in your area. As I've mentioned, you can use soft red for bread, for example, even if that is not done commercially. Bread from the soft wheats is just as good as from the hard wheats, we think. The latter, with more gluten in them, make bread more "efficiently"—at lower cost per pound of flour—so these are preferred by commercial bakeries.

There are dozens of varieties of wheat for each type, and new ones are developed routinely. Find out from neighboring farmers, the local feed store, or your local farm extension agent which varieties are recommended for your particular area.

Cereal grains are much more difficult to hybridize than corn, and results have not proven satisfactory so far. But hybrid wheats do show promise for the West, from Texas to Montana. Hybrid wheats will probably continue to be of interest mainly in this dryland wheat country, where yields on average are lower than in the soft red winter wheat area of the East. But where new, non-hybrid soft red wheats can attain yields of over 70 bushels per acre consistently, growers are hardly going to be interested in higher-priced hybrid seed.

Hybrid wheats do not always represent the greatest option. Hybrids increase the farmer's dependence on the seed producer, as I keep saying, because you can't save your own seed from hybrids. Responsibility for providing seed for our annual food supply falls into the hands of fewer and fewer people, which increases the risks in terms of biodiversity and food security. And as plant breeders have proved with cereal grains, it is possible and practical to improve yields, adaptability, and disease resistance without hybridization.

Newer wheats in both the eastern and western United States are shorter in height than older varieties. And the straw is stiffer. These wheats resist lodging better, even with high rates of nitrogen fertilizer. Organic growers may prefer the older, taller wheats, if they desire large quantities of straw for mulch or bedding. However, even the shorter wheats produce a good amount of straw.

Should you plant certified seed, germination-tested and weed-free? I have to answer yes, because that's the "right" way. If you are growing wheat commercially you'd be foolish not to use certified seed because, where profits are important, why be half-safe? The newer certified varieties are more disease resistant too, especially to Hessian fly and leaf rust. Some newer varieties ripen sooner than older varieties and make double-cropping soybeans or late vegetables after wheat practical as far north as the Corn Belt. When you buy certified seed, you know what you're getting.

But I also have to point out that, in a year when the wheat is generally of high quality (weighing close to 60 pounds per bushel), you can save money by going up to the grain elevator and buying a couple bushels, or whatever you need, of cleaned wheat (cleaned of weed seed), just as it comes in from the field. Invariably it will grow just fine, and though you probably won't know which variety you have, who cares? It is one that obviously came from your area.

Organic growers who want untreated seed won't have much of a problem getting some. Organic farming has come a long way, and many seed dealers now carry organic seed.

About half the seed wheat is not treated with chemicals anyway because many dealers want an "out." If the wheat doesn't sell as seed, they can at least sell it as feed if it's untreated.

The most economical way to "buy" seed is to save your own. If you harvest your crop with a combine that has good cleaning capacity, most of the weed seed and chaff will be separated out in harvesting. With older combines your wheat may have some weed seed, bugs, and chaff. Small amounts of grain can be winnowed clean by hand in front of a fan or in a stiff breeze. Or you can get a seed cleaner, which used to be standard equipment on every farm. Older cleaners are rapidly becoming collector's items, though you can still purchase them at farm auctions.

New seed cleaners are available from farm-supply houses, even by mail order. Nasco (Fort Atkinson, Wisconsin, 53538; or 1524 Princeton Ave., Modesto, California, 95352) is a reliable order house. Commercial models for commercial seedsmen are available in many brands, at much higher prices. If interested, you might write for the catalog from Burrows Equipment Co. (1316 Sherman Ave., Evanston, Illinois, 60204). But going on the Internet is the best way to learn what is going on in the whole arena of seed processing.

Growing Wheat

You can plant a small plot of wheat as easily and as simply as you would plant a small plot of lawn, because that's what wheat is, a grass. Work up a fine seedbed with rotary tiller, rake, disk, or harrow, broadcast the seed on the soil surface, then rake or harrow lightly to cover it. That's about all the work there is until harvesttime.

The time of planting winter wheat in the fall is important. Because of a pest, Hessian fly, you should wait until after the "fly date" in your area, usually around September 15, before planting. Ask any farmer in your area for the proper fly date after which Hessian fly is not active. Plant too early and you run the risk of an infestation. With new, Hessian-fly-resistant varieties, the risk is not great, but why plant too early anyway? You've got plenty of other things to do in the first half of September. If you don't, go fishing awhile. I've seen wheat planted as late as November 5 barely come up in the fall, and still make a moderate crop the next year. If you do plant early, or if the weather remains unusually warm late in the fall, your wheat may begin to "stool," or develop a stalk, which means it could more easily winter-kill later on. In some areas, like Kansas, farmers graze their wheat, which will control stooling. In spring, grazing can be continued until warm weather, with grain harvested from the same plants a couple of months later. This practice may cut yield a little, but often the early grazing is worth more than the increase in wheat yield that would be obtained by not grazing.

Wheat is not nearly as demanding of fertilizer as corn is. Keep your pH for wheat as close to 6.4 as you can and plant in well-drained soil, and half your growing problems are avoided. Wheat doesn't like acid soil and hates wet soil. Wheat will get enough extra fertility from what was applied to corn the year before, or what is left over when soybeans precede it. Commercial wheat farmers will disagree with that advice unless the price of fertilizer is high, but as I say often, the small grower is not after top yields. If you do fertilize wheat heavily, it is more prone to lodge or to be attacked by fungal diseases.

A typical low-nitrogen treatment for wheat on chemical farms might be 30 pounds of actual nitrogen per acre at planting and another 30 pounds top-dressed on the wheat in early spring. Organic growers can equal that application with several tons of manure per acre. In a fairly fertile soil that has been green-manured in the recent past and that grew soybeans the preceding year, that much manure contains all the extra nitrogen you need. You'll need to add phosphorus though. A 2-ton application per acre of finely ground rock phosphate every four years should be your standard program in a strictly organic regimen. I must confess, however, that I never use rock powders any more. I mulch with tree leaves, grass clippings, spoiled hay and manure and that is enough for me. I do not yearn to become the Michael Jordan of garden farming. Wheat does not seem to respond to applied potassium very much, and will get all it needs from mulches and manures in preceding years. Wood ashes are also a fair source of potassium.

Wheat provides a good example of what the agronomists mean by "balanced fertility." If you put 40 pounds of actual nitrogen per acre on wheat where there is not an adequate amount of phosphorus, you will fail to get a good crop, at least not a yield like you'd get by putting on 40 pounds of nitrogen *and* 40 pounds of phosphorus. (A balance does not necessarily mean the same amount: it just happens to work out that way with wheat.) Without the phosphorus, the wheat can't use the nitrogen. Likewise, supplying that much nitrogen and phosphorus, but not 15 to 20 pounds of potassium to balance it, can result in the crop lodging. Wheat needs the potassium for building stalks strong enough to handle the extra growth from the added nitrogen and phosphorus. But again, this is not so much a problem with organic fertilizers. Manure, for example, is a fairly well-balanced source of major and minor soil nutrients. With chemicals, a farmer can more easily get one nutrient out of whack with the others. At best, he's wasting money.

Sowing Wheat

Broadcasting, that is, spreading the seed on top of the ground, is a cheap way to plant the seed. Sometimes, though, birds will eat too much of the wheat if it is lying in plain sight on top of the ground. A grain drill will do a better and more precise job. Older grain drills are fairly easy to buy from dealers in used farm machinery. With a drill, the seed is placed at whatever depth you desire (2 inches is about right when the soil is dry, an inch or less when plenty of moisture is available) in rows spaced just a few inches apart. With a drill, you plant about 2 bushels per acre. Broadcasting requires a little more than that for a comparable stand because the seeds won't germinate as well. In broadcasting wheat, it is a good practice to go over the field with a cultipacker to press the seeds more firmly against the soil surface. The cultipacker will also cover many of the broadcasted seeds with a wee bit of soil, which is good. If you get a good rain right after the wheat is broadcast, the seeds will sprout better.

The ease and cheapness of broadcasting usually makes it worthwhile for the small grower. You can sow four acres easily by hand in a day with no more of a tool than a hand-cranked broadcaster, which costs less than a hundred bucks. Broadcasters used to be mostly canvas bags to hold the grain and a hand-cranked spinner to spread the seed on the ground affixed to the bottom of the bag. Now they are more often plastic containers. The container is filled with seed, the opening in the bottom set at the desired rate of seeding, and, with the whole apparatus slung over a shoulder, the sower walks at a steady pace, hand-cranking the spinner, which throws the grain out evenly in a swath on both sides of the sower. The opening through which the seed falls onto the spinner has numbers at the various notched settings, but whether you actually sow 2 bushels per acre at the "2" setting depends entirely on how fast you walk and how fast you turn the crank and what kind of seed you are sowing. Those numbers have to be calibrated to your pace and cranking speed, so to speak. Don't get too uptight about it. Just concentrate on walking at a steady pace and cranking at more or less the same speed. Then, if you know how big the area is you want to plant, you can ascertain pretty quickly if you are getting the proper amount of seed on. A little experience goes a long way. Better to put on too little seed the first time than too much. You can always go over the ground twice.

Walk straight across the field as you crank. If the seed falling on the ground seems to be about the rate of one seed per square inch, that is about right. Note how far out the seed is thrown from where you walk. Then on your return trip you'll know how far to walk from your first path so that you don't miss any ground and don't overlap either.

Larger broadcasters on the market operate on the power take-off shaft of a farm tractor and sow just about as accurately as drills. On very small plots, you can toss the seed on with your hand.

Before seeding, the ground should be worked, as already mentioned. In midwestern crop rotations, wheat is normally planted in the fall following soybeans, although other rotations may be satisfactory, too. However, if you do plant after soybeans, or any legume crop that fixes nitrogen in the soil, disk the ground in preparation for planting only lightly several times. Try not to turn up the old bean roots on which there may be attached nodules of nitrogen. Exposed to air, the nodules lose nitrogen that would otherwise be available for the wheat. Pulling a harrow behind the disk will help level the field, though the bean straw may plug up in the harrow too much. In the garden, the rotary tiller can be used to work the soil lightly after edible soybeans, or after potatoes or other vegetables before broadcasting the wheat.

After broadcasting, go over the field or plot lightly with cultipacker or rake. That will cover most of the wheat seed. Don't worry that some of the seed is not covered. Sowing at the 2-bushels-of-seed-per-acre rate, enough seed should germinate for a crop. The seed on top of the ground will sprout too, if rain falls after planting, as it usually does in the fall.

If you would like to use your wheat for the dual purpose of grain and grazing, the most practical way to do this on a small plot would be to let chickens peck at it late in the fall. Be sure to let the wheat grow for about a month before grazing it, so that roots develop well before the chickens peck the green blades away. A light grazing by sheep or a cow would be okay too, but if the ground is muddy, as it might be in late fall in the East and Midwest, the animals will trample and pack the soil too much.

A half-dozen chickens could be turned on a 20-by-60-foot patch of wheat in the spring to graze. The chickens' consumption will cut your total yield of grain, especially if you let them go on eating the wheat after it grows up and heads out, but you have to feed the chickens anyway, so this may be the cheapest strategy. But for wheat you grow for your own flour, it's best to pen the chickens away from it after it stools out.

Problems with Wheat

When wheat is heavy and high-yielding, wind may knock it flat on the ground, making harvest exceedingly difficult. This is called *lodging*. Shorter, stiffer-strawed wheats have solved the problem to some extent. But too much nitrogen, producing rank, succulent growth, may still cause lodging. A rich garden soil with plenty of humus will not need any added nitrogen. Lack of potassium, which is the nutrient that strengthens the stalks, can also cause lodging, but again, a garden soil that has been well mulched and/or manured in preceding years will have enough fertility to grow wheat.

Rusts and blights have been problems with wheat traditionally, but continual breeding of resistant varieties has held these diseases at bay. Smut, revealed by black heads of decayed wheat, used to dot wheat fields when I was a child. Normally it is not seen much today. (Today, if you mention smut, everyone thinks you're talking about that racy stuff you're supposed to pretend you've never looked at.)

"Take-all" is the common name for an old and serious disease that periodically rears its ugly head in the eastern Corn Belt. But don't get uptight about it. I have never experienced this disease, and I've lived for many years next to wheat fields. The disease is caused by a soil-borne fungus that grows best in wet conditions. Symptoms are stunting, premature ripening, and lodging. Roots rot and stem bases disintegrate. Stems often blacken near the crown, and then the plant breaks off at ground level. Losses can be total, hence the name "take-all."

Once infected, the wheat can't be saved by any means at man's disposal. But control is not difficult. The disease seems to strike wheat most frequently on poor soil, and is rarely seen on fertile soil. One control is to avoid planting wheat after wheat, proving again the preventative powers of crop rotation. Don't use straw for mulch from a field known to be infected with "take-all," as the fungus can be carried to another field in that way.

The cereal leaf beetle used to be serious and is trying to make a comeback, although no one knows why at this writing (in 2008). Many entomologists theorize that the mild winters of late have something to do with the upsurge, so if the temperature falls to 15 below next winter and freezes the ground down 8 inches or so, don't complain too much. I think that might get rid of the infernal Japanese beetle for a while too. The cereal leaf beetle's larvae, which look like little black slugs, like to attack the main new leaf of the wheat plant that starts to grow in the spring. That leaf is referred to as the "flag leaf," and its good health is central to the emergence of the seed head later on. Years ago this beetle was a serious threat, and its decline is one of the big stories of organic farming success, although rarely mentioned. Parasites of the beetle, mainly parasitic wasps, were introduced into affected fields and the problem soon diminished. Organic growers also have an approved spray now, called Entrust, from Dow Chemical.

Weed Control

The worst problem in raising wheat organically is weed control. Because wheat is customarily planted "solid" rather than in rows, you can't easily weed it, so without very good management, you can get too many weeds. Chemical farmers spray herbicides, which control most weeds in wheat fairly easily, except for a few new exotic weeds that appear to be immune.

Rotations help to avoid some weeds. In organic farming the crop before wheat should always be a row crop that has been cultivated intensively for weed control. That way you at least start off ahead of the weeds. Then, where wheat is sown in the fall after most weeds quit growing, the crop gets a good jump on weeds, makes a good stand, and is off and growing in the spring, choking out some of the weeds that try to come up later.

But don't plant solid-stand wheat in a field that has been full of weeds the preceding year unless you use herbicides. If you are organic and growing only a small plot, it's better to plant your wheat in rows and cultivate it like the Chinese do. American farmers may laugh at you, but the Chinese have forgotten more about raising food than we yet know.

In your rotation of crops, either in the field or in the garden, wheat is a good crop in which to plant clover for nitrogen fixation and as a green manure. The wheat in the spring is growing on a fairly well-cultivated soil surface. The clover seed falls on rather bare land, even though the wheat is growing there, and will sprout and grow readily. The wheat then acts as a "nurse" crop for the legume, which comes on to heavy growth after the wheat is harvested. More on that in chapter 11.

Since corn should be the first crop to follow the clover, the basis of your organic-grain rotation will be either wheat, clover, field corn, and back to wheat, or wheat, clover, sweet corn, back to wheat. Since it is good to follow corn with another nitrogen-fixing legume, soybeans, peas, snap beans, or lima beans are fine, making a rotation of wheat, clover, corn, beans, and back to wheat. Potatoes, wheat, clover, back to potatoes is an excellent rotation where potatoes are a main crop. A five-year garden rotation could be: wheat, clover, sweet corn, peas, and beans double-cropped to fall vegetables, tomatoes, then back to wheat. But almost any variation will work well if you maintain the basic wheat-legume-corn rotation and don't follow two vegetables of the same kind or family in successive years.

Harvesting Wheat

Winter wheat begins to ripen in the South about June 1, and the time of harvest moves northward from then into Canada, where it begins about August 1. Ripe wheat turns a flat yellow to almost a dead brownish red, depending on locality and variety. On most varieties, the heads crook over and point to the ground when the wheat is ripe. Pull a few heads, rub the kernels out in the palm of your hand, blow away the chaff like a pro, and chew a few grains. If crunchy hard, the grain is ripe; if at all chewy soft, it is not yet ready. Moisture content should be 12 to 13 percent before grain can be safely stored.

If you want to be real scientific about it all, you can use a moisture tester to determine the fitness of your grain for harvest and storage. Grain elevators all have the testers. Take a half-bucket sample for testing. But for the small grower, a few bushels of wheat can be spread out in a pickup bed or on a clean floor to dry sufficiently, if you have any doubts when you harvest.

The best and easiest way to harvest your grain if you have more than half an acre is to hire a custom combiner to do it. He'll look pretty silly pulling his monstrous machine into your small patch—in fact he may refuse to come for anything less than ten acres or unless you pay him extra. But a nearby farmer who owns a combine can usually be persuaded to harvest a little patch of grain, after he finishes his own.

On small plots, you can harvest by hand. We have, on occasion, made a sort of party out of the harvest. Everyone has fun and the children learn something.

For hand threshing, the grain is cut, tied into bundles and shocked first. For this operation, you do not wait until the wheat is completely ripe, but cut it a little green, when the wheat is still in the doughy stage. The stalks will be yellow, but you will still discern streaks of green in some of them.

Scything wheat.

Carol, Jerry, and Jenny Logsdon winnowing wheat with a box window fan.

Cut the wheat with a scythe, leaving perhaps a 3- or 4-inch stubble. Cut about 2 feet of standing wheat at a time, measuring from the outer edge of the uncut stand. You will just have to practice until you figure out what, for you, is a natural, easy swing of the scythe. You should use a grain cradle if you know where to find one outside of museums. The grain cradle is a scythe with a set of wooden tines above the blade that catch the wheat as the blade severs it on your forward swing. At the farthest reach of the cradle's forward swing, the wheat on the tines falls into a neat pile, ready to be tied into a bundle.

But grain cradles are antiques now, so you'll probably have to use a regular scythe. I can't tell you how to swing a scythe because I am not accomplished at it myself. If you grab the handles and swing the blade into the standing wheat ahead of you, the scythe will sort of teach you how to fall into the proper motion and rhythm. The only advice I can give with printed words is to try not to take too big a "bite" of wheat at each cut. Instead, cut a narrow swath, letting the blade of the scythe slide against the wheat stalks at a more or less 45-degree angle as you swing through them, rather than whacking into them squarely from a 90-degree angle. And take your time. Remember the old German farmers' saying: "Slow and steady goes far into the day." Sharpening is half the battle. Another old saying goes like this: "If you have four days to cut down a tree with an axe, spend two days sharpening it."

You can, of course, cut the wheat with a sickle-bar mower, and then rake up bundles from the cut wheat with a leaf rake. Tying the bundles can be a tedious task unless you've invited friends over for a picnic and/or party. Then, while the men try to show their prowess with the scythe, the women and children can tie bundles. We use baling twine to tie the bundles. In the old days, farmers knew how to tie bundles using strands of the straw itself. I've never mastered the trick, though I've been shown how on several occasions. Basically, you take two or three long wheat stems together, wrap them around the center of the bundle you wish to tie, and then twist the tie strands around each other against the bundle. A bundle should measure about 8 inches in diameter in the middle where it's tied.

When the bundles are all bound, you either set them up into shocks, or stack them in the barn out of the rain. About fourteen bundles make a good shock. Over the top of the shock one or two bundles can be flattened and bent to make a cap to deflect rain from the grain.

Grain will ripen in the shock and remain virtually unharmed by rain for about a month. If you have only a few shocks, as I'm assuming, it's best to get them into the barn as soon as possible after the grain is ripe—or even before—since rain is always somewhat harmful to the grain.

Hand Threshing

Once the wheat ripens in the shock—about three weeks out in the sunshine, longer in rainy weather, you can thresh out the grain. Lay out a large clean cloth (an old bedsheet is fine) on a hard surface, such as a sidewalk or patio or wood floor. Lay a bundle of wheat on the sheet, and whack the daylights out of it with an old broom handle, plastic toy bat, or other appropriate club. A friend of ours uses a length of rubber hose. You won't have to strike the wheat heads hard, as the grain will shatter out quite easily onto the sheet. Not every grain will fall out. Throw the bundles already flailed to the chickens. They'll pick out any grains you missed, and the straw becomes their bedding.

Each bundle will have a cup or two of wheat in it. After you have flailed several bundles, pull the corners of the sheet together and dump the grain, chaff, and bits of straw into a bucket for later winnowing. Winnowing means separating out the chaff with some kind of forced air. If a strong breeze is blowing, you can pour the contents of your bucket into a second bucket from a height of 3 or 4 feet and the wind will blow away much of the foreign matter. I use a big window fan, which gives a steadier and more reliable blow.

In front of the fan, I pour the grain from one bucket to the other six to ten times before all the chaff and straw winnows out. The heavier grain falls nearly straight down into the receiving bucket and the fan blows all lighter material beyond the bucket's rim. (Do the job outside!) You won't get the grain perfectly clean unless you really work at it, but no matter. The few bits of wheat hull that persist in sticking with the grain won't hurt a thing. It all grinds up to flour and makes the fiber content that much better. But the longer you winnow, the smarter you get about it, noting that you can vary the height of your pour or the distance away from the fan to get rid of nearly all the foreign matter. If you want grain winnowed completely clean, you can use a seed cleaner, described earlier.

There are other harvesting alternatives for small plot growers. Small threshers and hullers are still very much used by the seed-processing trade, but the machines are too expensive to be practical for most garden farmers. You can find plenty of references to all kinds of models, expensive and not so expensive, on the Internet or in some farm-supply catalogs.

A friend writes recently that he was able to find an older, tractor-powered combine, still in good shape, to pull behind his small tractor. If at all possible, get the service and parts manuals for the combine from the person you buy it from. Watch for antique farm machinery shows in your area where these manuals are often for sale. Such shows are the best place to hunt for old combines and other antiquated machinery that is just right for small-scale grain growers.

Leaf and twig shredders will do a crude job of threshing grain too. With modifications, they might do better. Try taking out every other blade. Or remove the screen completely. Gear the motor down so it will run the shredder slower. I've even threshed wheat with my lawn mower, with a sheet of plywood underneath and a board to one side to block the grain from being scattered too far by the blade. If you have a horse, you might want to try threshing by trampling as was commonly done in earlier days. Let the horse walk on the bundles on a clean wood floor. Grain is still threshed this way in many countries. I wonder, though, what they do when the horse decides it needs to relieve itself.

Storing Wheat

It's easy enough to tell you to store grain in insect-proof containers, and that's easy enough advice to follow. But keeping the moths of grain beetles or weevils out of your dried wheat is not the main problem. Sometimes the weevil eggs or larvae may already be in the grain when it is harvested or when it is binned before shipment or packaging. The best way to store your grain for grinding into flour is to keep it refrigerated—in a freezer is best. You can use plastic containers to hold the grain. You can't store very much that way, of course, unless you have an extra freezer just for that purpose. But half a bushel will serve a lot of breakfasts.

Dry ice is another method advised for "fumigating" weevils out of grain. It's considered an organically safe way to protect grain, but I wonder if the carbon dioxide from dry ice is much safer than malathion, one of the less toxic chemicals that is quite effective in weevil control. A container of dry ice not properly handled can blow up too. Above all, don't use dry ice in a glass container. Use a metal, five-gallon can, spreading two ounces of crushed dry ice on the bottom and putting the wheat on top of the ice. Allow sufficient time—about half an hour—for the dry ice to evaporate before placing the lid on the can. If the can starts to bulge, remove the lid cautiously for a couple of minutes and then replace it. I really don't recommend it myself and have never resorted to any of these methods.

My weevil control starts before the grain is put in the bin. Clean the floor and walls thoroughly. If rats or mice have gnawed holes in the bin, patch them over with pieces of tin secured with roofing nails. I then spray the bin with Raid or a similar insecticide and let it stand empty for a few days. For organic purists, dust the bin with a pyrethrum preparation or with diatomaceous earth. The wheat is then treated with diatomaceous earth too, at the rate of one measuring cup worth to 25 pounds of grain. Thorough mixing is necessary.

Diatomaceous earth is the fossilized shells of tiny one-celled plants called diatoms. Deposits of these fossils are quarried, milled, finely ground, and screened into a talc-like powder that can dehydrate and kill insects. It is considered harmless to humans, plants, and animals unless inhaled in very large amounts. There is some controversy over the practicality of diatomaceous earth as an insecticide. Even the USDA reports are conflicting. However, the USDA recommends the material for weevil control right along with recommendations for chemicals—something that the USDA rarely does with organic controls—so it must be effective. I am not all that precious. I want the weevils out of there for sure, so I treat the bin with an insecticide before storing wheat in it.

It is better to store wheat in wooden bins so that the grain continues to dry out if it has not completely done so before. Wood is more porous and absorbent than metal. If you store your wheat in steel drums, it is better protected from rodents, but it needs to be very dry. My wheat bin is made of plywood and has stood me well. If you use a steel barrel, obviously don't use one that had oil or some toxic material in it. Even after cleaning, the barrel will carry the odor of what was previously in it, which deters weevils and other bugs perhaps, but may hurt the taste of the wheat.

Eating Wheat

I don't intend to describe the hundreds of ways wheat can be turned into human food. There are many good whole-grain cookbooks, and there are a number of wheat-based recipes at the end of this chapter. Using whole grains is not a matter of availability of recipes. There are tons of them. For those of you who, like myself, develop a severe case of laziness at the thought of cooking anything complicated, there are ways that require very little time or skill.

I grind up a cup or two of wheat in the blender, or used to when I was young and full of ginger. Then I sprinkle it over breakfast cereals, like oatmeal or dry cereal, or eat it alone with cream and honey. It has a husky, nutty flavor that I like. If you don't, there's an easy way to "cook" wheat. Pop it in a shallow pan like you would popcorn. You don't need a cover on the pan as the popping wheat won't fly about. You don't need oil in the pan either, though you must keep the pan agitated so the wheat won't burn. The kernels "pop" only slightly, but the wheat is then crunchy and easy to chew and quite delicious, salted and/or buttered. It makes an excellent and healthful party snack food.

If you then grind the popped wheat in the blender and eat it with cream and honey, the roasted-nut flavor is even more pronounced and more to most peoples' liking than the raw taste. The popped wheat can be cooked for a hot cereal, too. Cooking wheat otherwise involves an overnight soaking.

To make pancakes, grind the raw wheat and sift it to make it a little finer. Then mix it into your batter and then into the frying pan as usual. Sorry, Pillsbury, you just can't equal the result. And it's so easy even I can do it.

My wife makes wheat cookies and a marvelous pie crust with the wheat, and of course makes bread with it, but she uses some more refined purchased flour with it (thank you, Pillsbury) that makes the bread a little lighter. Every cookbook has many bread recipes.

Feeding Wheat to Livestock

Mother hen up at the barn was hatching a clutch of eggs. Two chicks were up and about while the rest of the eggs remained unhatched. Mother didn't want to leave the eggs, but the two chicks were hungry and needed Momma to forage a little food for them. Rather than run to town for some prepared chick feed, I took a saucer of ground wheat up and sat it in front of mother's nest. As soon as I backed away she began to cluck like crazy, grabbing a pinch of ground grain in her beak and dropping it under her for the chicks. She didn't even have to get up. In a few moments, the chicks came out from under her wings and ate the wheat with obvious gusto.

Except for tiny chicks, wheat kernels are small enough for farm animals to eat whole, but if you want to grind your grains, farm-equipment dealers all sell hammer mills for that purpose. Often old ones sell cheap at farm auctions or are left rusting away in the back of barns. Older ones are all powered by tractors via the power take-off shaft or by way of belt and pulley; new ones more often by electric motor. Hogs should not be fed whole wheat except in small portions. Whole wheat hogged down by a hungry pig doesn't get chewed properly, and in larger amounts the whole grains may swell in the pig's gut and cause problems. Unlike sheep and cows, hogs have only one stomach. The real point here is to feed grain only sparingly at any one time to animals. Lambs especially can get what is called "overeating disease" from a diet heavy with grain.

Wheat has more protein in it than corn but less carbohydrates. It takes about one-and-one-third times as much wheat as corn to produce the same increase in weight, which is another way of saying that corn is more fattening. And since wheat is worth more as human food than as animal feed, it is not widely used to fatten animals. But it can be. Whole wheat and corn mixed half and half makes a good scratch feed for chickens.

I've mentioned that I like to feed chickens wheat in the bundle so the chickens, instead of me, do the threshing. The drawback to this method is that, in the bundle, the grain is more vulnerable to rodent and weevil attack than when the grain is threshed and stored in a rodent-proof bin. My solution is to feed all the wheat bundles from harvesttime until the corn is ready, before the weevils and rodents have time to do much harm. Then feed corn through the winter.

In milled animal feed, oats are probably more efficient than wheat in combination with corn. But for feeding whole, most animals (except horses) and especially chickens prefer wheat to oats. Chickens will eat whole oats if there is nothing else around, but the tight hulls on the kernels are not much to their liking.

Straw

Wheat is an extremely versatile crop, food for both animals and humans and able to be grazed too. But the straw also provides another important commodity after the grain is harvested. Straw is commercially valuable sold in bales. It is used mostly as livestock bedding, but also in the production of some paper products.

No organic-minded farmer wants to sell off his straw unless he desperately needs the money. Straw is worth more to him as organic matter incorporated back into the soil than the money in the cash market. If he keeps livestock, the straw goes for bedding and then back on the field as manure. If he doesn't keep livestock, he should incorporate the straw back into the soil directly. Otherwise he has to use the money he gets from selling the straw to buy fertilizer he'll need to replace the loss.

If you need larger amounts of straw and have harvested a large field of wheat with a combine, rake up all the straw into windrows as you would for making hay, and then bale it. If you do not have access to a baler, you can load the straw on a wagon or truck with pitchforks and haul it into your barn and stack it there out of the rain for use as bedding. Or you could make a strawstack outside, as in the old days.

I cannot think of strawstacks without getting sentimental, but the strawstack's value goes far beyond sentiment. The strawstack was the real symbol of agriculture, even more so than the pitchfork, from about 1880 to 1930, the period of agricultural dominance in America. And, I think, with good reason. We lost a lot more than cheap straw when the stacks disappeared from the landscape. Each of those stacks represented a commitment by the local community to the local community. I remember vividly the threshing rig that blew the threshed straw into those big, round-topped stacks. You needed at least a dozen farmers—the threshing ring—to use the machine efficiently. One man tended the steam engine that ran the thresher, two handled the threshed grain, one or two "built" the stack, and at least six to eight others loaded and hauled bundles to the thresher and fed them into the threshing cylinder. Wives and daughters prepared those huge thresher dinners. My mother liked to recall sarcastically (she did not enjoy preparing those huge meals) that the "womenfolk" waited around after the meal to hear the men belch. The louder the belching, the better the food, she joked. There had been at least as many women involved in preparing the food as men involved in harvesting the grain. Not all the women who came to help with the food did my mother remember fondly, and among the men there were individuals who despised each other. But all of them knew that they depended on each other for survival. Every strawstack was an expression of faith between farmers.

Happenstance allowed me to work as part of a threshing ring in my twenties, long after threshing had died out as the normal harvesting method. It just so happened that a group of farmers in Minnesota where I lived in the 1950s were "backward" enough to save money by sticking to their old thresher rather than all of them buying individual combines like everyone else was doing except the Amish. I loved those threshing days. I loved the look of all those rolling acres of wheat and oat shocks; the wagons and teams of horses hauling the bundles to the thresher, men bantering with each other, men uncommonly proud of their horses hauling the bundles to the thresher, or their skill at loading, or their strength. I remember women clucking and gossiping over their cooking, the long tables of food, huge pots and trenchers of it. That was food you could not buy with all the gold in the world.

The strawstacks were rallying points on each individual farm, too, a place where life flocked and frolicked all through the rural winter, or a place for courting by the light of the summer moon. The barnyard around a strawstack was always very alive. The farmer had to be there every day to take great gobs of straw inside the barn for bedding. His children tumbled and slid down the stack, though not always with father's approval. The cows snuggled in close around the stack, wearing the straw off where they rubbed until the stack resembled a huge mushroom with a thick stem and overhanging cap. Against the "stem," under the overhanging straw, the animals could huddle, snug from the snow and rain. Sometimes, a cow or horse would chew on the greener straws, eating—literally—a hole into the stack big enough to accommodate half its body. There were many stories of stacks collapsing on burrowing animals and smothering them.

The hogs had their own stack. A frame of logs or rails was first constructed and the straw then heaped over the frame. The room thus created inside the frame was warm and cozy enough for piglets even in the coldest weather.

All winter long, life teemed at all levels around the stacks. Manure droppings were covered quickly by fresh straw, and the composting process began in the dead of winter. Livestock could lie on the straw-covered manure pack and stay warm because the bacteria and microorganisms were already heating up the manure in their work of decomposition.

Chickens fluttered around the stack too, seeking grains of wheat still in the straw or a place to lay eggs. Pigs nosed about, eating the undigested grain from the manure of the cows. There was a chorus of mooing, neighing, squealing, clucking, and laughing, presided over by a farmer who, if he thought no other human was about, might very well be singing.

A year after a stack was gone, all that was left where it stood was a storehouse of nitrogen, phosphorus, potassium, and humus. Certainly, as my grandfather often proved, there was no better place to plant watermelons, even in the North where watermelons weren't supposed to grow so well.

But the strawstacks are mostly gone now except on some farms of the stricter Amish sects. And on too many farms, the animals are gone too. In winter, the barn lot is an empty, windswept, forlornly cold place to be. The old barns stand as silent as mausoleums, replaced by modern stinking, chemicalized, electrified, concrete-clad "systems" where animals are crammed together like fistfuls of maggots.

The strawstacks have disappeared, the barns stand empty, and the farmers hope to go to Florida in the winter. Their children have gone away too. You will find them sitting stolid and impassive before the television set, watching Disney films that show them how children used to learn about life: caring for animals and sliding down strawstacks. No longer do children have the blissful independence to relax like Little Boy Blue, "under the haystack, fast asleep."

Toward a Perennial Wheat

At his Prairie Institute near Salina, Kansas, an intrepid plant scientist, Wes Jackson, is carrying on what are surely the most exciting long-term experiments in the world of grains. If his years of effort come to fruition—and there is no reason why they should not, given lots of time—we humans will have available a grain, or possibly several different kinds of grains, that will no longer have to be planted every year. They will come up every spring like lawn grass, and harvested like wheat, without any annual cultivation. They will be, in other words, perennials.

One of the promising wild grains Wes and his staff are growing is intermediate wheatgrass, a sort of wild wheat. (They also have experimented with an old wild species of wheat called emmer, which is itself drawing interest from some garden farmers.) He is both crossing wheatgrass with regular wheat and selecting promising plants within the wild wheatgrass family in hopes of finding a perennial that will grow enough seed of enough size to be economically practical. The work is arduous and time-consuming, but already selections are producing enough sizable seed to experiment with for bread making and other prepared foods. Since wheatgrass varieties are being bred also for grazing at various Great Plains universities, think what this could mean eventually: a nation of golden grain from sea to shining sea that we could make bread and meat from without turning over one shovelful of dirt.

Coincidentally, while these experiments are in progress, wheatgrass has also become the rage for juicing into a healthful drink. (A good website for information on how to grow and use wheatgrass for juicing is ShirleysWellnessCafe.com.) Amazing claims are being made for the health benefits of wheatgrass juice (and also for juiced barley grass). Sprouting wheatgrass seeds is about the same procedure as for any other sprouts, although the Web site cited gives some details not often contained in other directions. But when the sprouted plants are about eight days old, grown in trays, the plants are juiced instead of eaten as a salad.

Seeing all these developments, it becomes amazingly clear that the growing and eating of grains on a small scale is only in its infancy. Wouldn't it be awesome if in the future we grew garden grains to drink as well as to grind into flour? And I don't mean alcoholic drink, although that too is an eminent possibility.

Wheat Recipes

These recipes using wheat are just a few examples to inspire your imagination.

Crusty Oven-Baked Fish

2 pounds haddock or flounder fillets
½ cup wheat germ
½ cup peanut flour (raw peanuts ground in a blender or nut grinder)
¼ cup sesame seeds
½ cup bran flakes or whole-grain bread crumbs
1 teaspoon salt
½ teaspoon black pepper
½ teaspoon oregano
½ teaspoon marjoram
½ teaspoon paprika
½ teaspoon garlic powder
1 egg, beaten
½ cup oil
½ cup lemon juice

• Preheat oven to 400°F.
• Rinse fish. Cut into portions and leave to drain.
• Combine all dry ingredients to make crumb mixture and set aside.
• Combine egg, oil, and lemon juice in a blender or use an egg beater to obtain an emulsion.
• Dip portions of fish into egg dip and then into crumb mixture.
• Lay on a shallow baking pan that has been lightly oiled. Bake in the oven approximately 20 minutes, until tender.

Yield: 6 to 8 servings

Apple Walnut Loaf

½ cup oil
1 cup honey
2 large eggs
1 teaspoon vanilla
4 tablespoons yogurt
1 teaspoon baking soda

½ teaspoon salt
2 cups whole wheat flour
1 cup walnuts, coarsely chopped
1 medium apple, unpeeled

- In a large bowl, mix together well the oil, honey, eggs, vanilla, and yogurt.
- Add the soda and salt and mix very well.
- Stir in the flour, until thoroughly mixed. Stir in the walnuts.
- Wash and core, but do not peel the apple. Cut the apple into largish chunks and add it to the batter.
- Pour batter into two well-greased 7½-inch loaf pans. Starting in a cold oven, bake at about 350°F for about 50 minutes.

No-Knead Whole Wheat Bread from County Cork, Ireland (Myrtle Allen Bread)

4 teaspoons dry yeast
⅔ cup lukewarm water
2 teaspoons honey
5 cups whole wheat flour
3 tablespoons molasses (unsulphured)
⅔ cup lukewarm water
½ tablespoon salt
⅔ cup wheat germ
1⅔ cups lukewarm water
½ tablespoon butter
1 tablespoon unhulled sesame seeds

- Sprinkle yeast over lukewarm water. Add 2 teaspoons honey. Leave to "work" while preparing the dough.
- Warm whole wheat flour by placing it in a 250°F oven for about 20 minutes.
- Combine molasses with ⅔ cup lukewarm water.
- Combine yeast mixture with molasses mixture. Stir this into the warmed flour, then add the salt and wheat germ and finally the 1⅔ cups lukewarm water. The dough will be sticky.

- Butter a loaf pan (9¼ × 5¼ × 2¾ inches), taking care to grease the corners of the pan well. Turn the dough into the pan. No kneading is necessary. Smooth dough in pan with a spatula that has been held under cold water to prevent stickiness. Sprinkle sesame seeds over top of loaf. Leave to rise to top of pan in warm, draft-free place. Meanwhile preheat oven to 400°F.
- Bake in oven for 30 to 40 minutes, or until crust is brown and sides of loaf are firm and crusty. Set pan on a rack to cool for about 10 minutes, then remove loaf from pan and cool completely on rack before slicing.

Cheese Wheat Germ Biscuits

3 tablespoons oil
¾ cup yogurt
¼ pound sharp Cheddar cheese
½ teaspoon salt
¾ teaspoon baking soda
¼ cup raw wheat germ
1¼ cups whole wheat flour

- Preheat oven to 400°F.
- Mix the oil and yogurt together in mixing bowl.
- Grate the cheddar cheese into the same bowl. Add the salt and mix.
- Mix the baking soda in very well.
- Add the wheat germ and flour and mix both in until well distributed.
- Pat out dough and cut into ¾-inch-thick rounds. Place on oiled baking sheet and bake in oven for 20 to 25 minutes.

Yield: 8 biscuits

Beef and Wheat Berry Casserole

1 cup whole wheat berries
3 cups water

½ teaspoon salt
1 pound ground beef
1 onion, chopped
5 tablespoons oil
⅓ cup whole wheat flour
2 teaspoons parsley, chopped
½ teaspoon sweet basil
½ teaspoon salt
4 medium-sized potatoes, cooked and mashed

- Preheat oven to 350°F.
- In a saucepan that has a very tight-fitting lid, combine wheat berries, water, and salt. Bring to a boil, cover, and remove from heat. Wrap pot in newspapers or a heavy woolen blanket and allow to stand overnight in a warm place. Drain and reserve liquid.
- Sauté the beef and onions in 2 tablespoons of oil and set aside. Drain and discard excess fat.
- Heat remaining 3 tablespoons oil in skillet, stir in whole wheat flour, and cook for a minute or so. Then add 1½ cups of the reserved wheat berry liquid and cook, stirring constantly, until mixture is thick and smooth. Add herbs and salt.
- Butter a 2½-quart casserole and put the wheat berries in the bottom, then a layer of meat covered with half of the sauce. Top casserole with mashed potatoes and pour remaining sauce over them.
- Bake in oven for about 30 minutes, or until casserole is hot and bubbling.

Yield: 8 servings

CHAPTER FOUR

The Sorghum Family

When it comes to the feed grains, corn gets all the glory. Point out that grain sorghum is almost equal to corn in nutrient value and that it will outproduce corn in dry climates, and in fact, will grow in dry climates where corn won't, and you are met with silent disbelief. Point out furthermore that grain sorghum makes a fine flour for human diets and the silent disbelief may turn to not-so-silent snickers. A great many people have never even heard of grain sorghum.

If you are driving down a highway in Texas and see a field of what looks like knee-high corn, only the leaves are narrower, that's grain sorghum that has not headed out yet. If you spot a field anywhere from the Corn Belt to Louisiana that looks like the tallest corn you've ever seen, but that doesn't have any ears—only clusters of small seeds on top—that's sweet sorghum. And if you notice a field of cornlike plants with bushy, whiskbroom tops where the tassels should be, that's broomcorn. All three plants belong to the sorghum family, and all three can have a place on your homestead, or in your garden.

Grain Sorghum

More often called milo, grain sorghum is a major feed grain grown in Texas, Oklahoma, and parts of the dry Great Plains all the way to California. Grain sorghum grows well throughout the South and competes with corn in some parts of the Corn Belt, though it certainly never takes the place of corn in humid climates.

Grain sorghum reaches a height of about 4½ feet with the seed clusters forming at the top of the stalk. In commercial plantings, the grain is harvested with a combine about like wheat. Yields

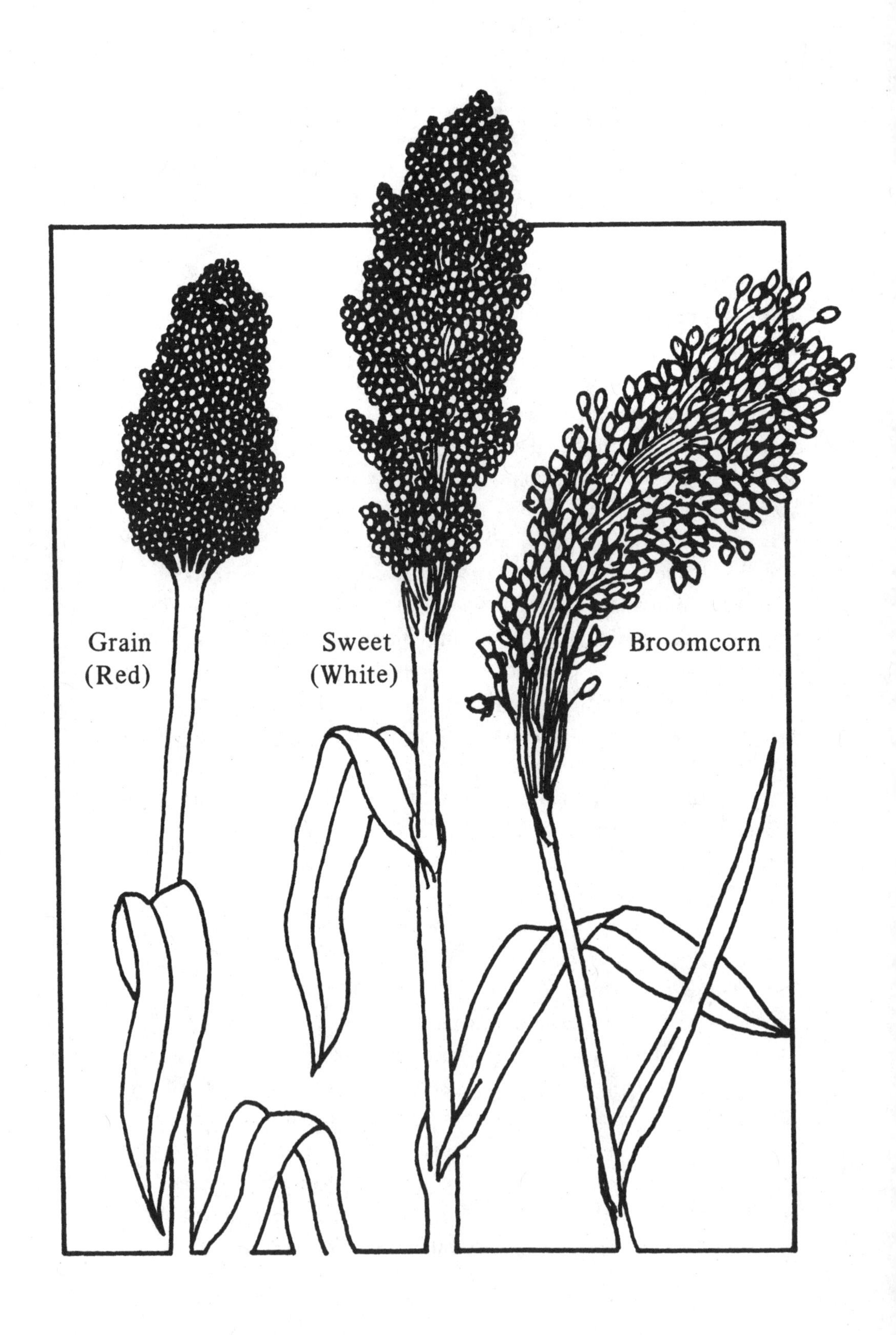
Grain
(Red)
Sweet
(White)
Broomcorn

are comparable to corn, about 100 to 180 bushels per acre on unirrigated land, depending on fertility. But where corn will only yield 80 bushels per acre in dryland western growing conditions, sorghum can yield 100 bushels.

Grain sorghum seed is brownish with yellow and red coloration mixed in. The browner the seed, the more tannin is present, generally speaking, and the lighter the seed, the less tannin. The higher the tannin content, the less palatability the grain has for cattle and presumably for human consumption. Lighter-colored varieties, especially those labeled "yellow endosperm" varieties, are the most palatable. Also, tannin content seems to be related to crude protein digestibility: the higher the former, the lower the latter. So although grain sorghum usually contains more crude protein than conventional corn, less of it is digestible. All of which would seem to direct you to selecting yellow endosperm varieties, especially if you are a homesteader looking for alternate kinds of whole grains for table use.

But if birds are a problem where you live, that choice may be debatable. Birds are the proverbial fly in the ointment to growing yellow endosperm sorghum because they love the stuff. Out West, in a one-thousand-acre sorghum field, the birds might wipe out several acres worth and no one would miss it much. But in Georgia, or Ohio, or similar areas, where birds are more numerous and fields smaller, raising the grain can be a mighty poor business. When a friend of mine grew a couple of rows of yellow endosperm grain sorghum in his garden for flour, the birds literally wiped out all of the clusters he didn't cover with paper bags.

So plant breeders, who are certainly a resourceful lot, have developed bird-resistant varieties. It was noted that the higher the tannin content, the less birds ate the grain, at least if there was anything else around to eat. So breeders developed grain sorghums that have a high tannin content when the grain is in the milk stage—which is when the birds really like to gobble it—but in which tannin content decreases at maturity, for better palatability. These sorghums can be planted even in the Lake Erie region where red-winged blackbirds can descend like a plague of locusts on farmers' fields.

The palatability of bird-resistant sorghums evidently affects only livestock. Chickens relish grain sorghum of any kind, and I can vouch for that. I have saved seed, even from sweet sorghum, which is not considered a feed grain, and the chickens loved it.

Grain sorghum seed certainly doesn't look like corn seed, being smaller than peppercorns and roundish. But when you plant your grain sorghum, think corn. Both are raised about the same way, though there's more variability in sorghum culture than with corn. Sorghum has been planted successfully in everything from 7-inch rows to 40-inch rows. A happy medium, 30-inch rows, is your best bet. In the row, you should maintain no more than three plants per running foot, and in dry climates less than that. The amount of seed farmers plant per acre varies all over the place too, depending on climate, row spacings, and fertility. Some farmers in dry areas plant less than 2 pounds per acre and in humid areas as much as 16 pounds. Again, strike a happy medium: 8 pounds is plenty. You don't want to plant sorghum too thickly or you'll have thin, weak stalks that lodge.

Plant grain sorghum about ten days after the proper corn-planting dates in your area. Because sorghum can be planted later than corn, southern farmers can plant it after harvesting wheat in June and so get two crops from the same field in the same year. Also, the later planting date of sorghum gives the small farmer or homesteader greater flexibility. If your corn planting gets delayed by bad weather, or your old equipment (or primitive hand-tool methods) is too slow to get the corn you need out on time, you can plant the rest of the acreage to grain sorghum. For instance, if you have only weekends for homestead work and it rains on two weekends in May so that you are facing June only half finished, grain sorghum to the rescue!

Cultivate for weeds as you would in corn, which means as soon as possible after planting. Sorghum looks more like grass when it first comes up, and you'll probably have it growing too thickly, so if your early cultivation buries a few of the plants, don't worry about it.

Diseases

I always dread writing about plant diseases because most of the time small-scale growers won't have a major problem, and, if they do, by the time they find out what's wrong, it's too late to do anything. And it is so easy to misdiagnose diseases. The standard advice might be useful for commercial, veteran growers, but they already know what's wrong and consult university experts to find out what to do about it. University experts don't always know, either. But if I don't mention these diseases, I will be accused of not giving complete information. In farming there is no such thing as complete information.

Grain sorghum diseases include bacterial leaf spot, bacterial streak, anthracnose, gray leaf spot, helminthosporium blight (don't try to say that one during Happy Hour), rough spot, rust, sooty strips, target spot, and zonate leaf spot. In the drier regions west of the Mississippi, such leaf diseases are seldom a problem of major proportions. In humid areas, they can cause trouble in a wet year. Crop rotation, sanitation, and the use of resistant varieties are the best ways to avoid bacterial and fungal diseases, as is true of most grains most of the time.

Three kinds of downy mildew attack sorghum. Two of them, popularly called "crazy-top" and "green-ear," are rare and of little concern to the homesteader. The third kind of downy mildew, however, has caused problems along the Texas Gulf Coast, has spread throughout the South, and has been discovered as far north as Indiana. Diseased seedlings show a white "down" on the underside of leaves. The down releases more spores, which cause lesions on plants. The leaves first become striped, then shred.

Fortunately, hybrids are available with a high degree of resistance, especially for growing in the humid South. Also, avoid growing Sudan grass or sorghum-Sudan hybrid grasses where downy mildew is a problem. These close relatives of sorghum can become heavily diseased and contaminate the soil for years.

Maize dwarf mosaic (MDM) strikes grain sorghum worse than it does corn, at least when sorghum is grown in the proximity of Johnson grass. Johnson grass serves as a reservoir for the virus, especially over winter. Then aphids feed on the Johnson grass and carry the virus to sorghum and corn. A typical symptom on sorghum is a light- and dark-green mottling of the leaves, discoloration of the heads, and stunted growth.

Eradicating Johnson grass would eliminate much of the problem, but that seems to be an impossibility. Cut down a Johnson grass shoot, and fourteen come up to take its place, seemingly overnight. Cut two and twenty-eight appear. A really energetic cutter can create a jungle in two days. And the stuff grows so fast you can't afford to leave a tractor in the middle of a field overnight. You might not be able to get it out until winter.

I exaggerate, of course, but not as much as you might think if you have never tried to negotiate with Johnson grass.

No sorghum hybrid is completely resistant to MDM, but some varieties are quite tolerant of it. Unless weather conditions turn unusually cool, an infection doesn't seem to make much headway. Hot weather seems to hold it in check.

A number of stalk and root rots damage grain sorghum: charcoal rot, fusarium stalk rot, red stalk rot, and some others of minor significance. The only control is to use recommended resistant varieties sold in your area. But remember that nutritional problems sometimes look like diseases. Iron deficiency in sorghum is a good example. It causes leaves to turn yellow with dark green veins and leaf tips to turn white. That's not very helpful, though, because any number of diseases also exhibit those symptoms.

If planted too early in cold, wet soil, sorghum seed is prone to rot, more so than corn. Nonorganic growers treat seed with a fungicide, but the best defense is to delay planting until the soil has warmed properly. One rule of thumb says to plant when the soil surface (2-inch depth) at noon registers at least 70°F.

Harvesting Grain Sorghum

Harvesting sorghum by hand is easier than harvesting wheat. When the seed heads are ripe and dry, the grain comes out of them easily enough; in fact, it will shatter out if left in the field very long after ripening. Go down the row after the seeds are hard, but not dead, falling-off ripe, and cut the seed heads off with about a foot of stalk. Pretend you are cutting a bouquet of seed clusters for display in a large vase or urn, which, by the way, is an excellent thing to do with the seed clusters. The brown-red-yellow seeds make an attractive fall table decoration. Tie the stalk heads together into bundles and hang in the barn, or spread no more than two or three bundles deep in a clean, out-of-the-way corner. Hanging or laid out, the clusters can dry until you need them.

In the North, grain sorghum grown commercially must often be harvested before it is completely mature, then artificially dried, because of the shortness of the season. If you allow the crop to stand in the field and the fall is wet, you might have trouble getting it harvested at all before it molds. Therefore, on the small homestead, cutting the heads before they are completely ripe and hanging the bundles in a dry barn is a doubly good idea.

Larger plots of grain sorghum will need to be harvested with a grain combine unless you have a large number of teenagers around with nothing to do between baseball and football seasons and can make them believe they are not harvesting milo, but playing for a championship in an up-and-coming new sport. A quarter of an acre seems to be the most to try to harvest by hand unless you are a real glutton for work. But sorghum will not be one of your "main grains," and a little will go a long way on a homestead. Since the grain will yield 100 bushels to the acre with a little luck, you ought to get 25 bushels from a quarter-acre, or 12 bushels from an eighth of an acre, or at least 6 from a sixteenth of an acre. Six bushels is certainly adequate for a small homestead. Some for you and some for the chickens. After chewing on the stub of my pencil for fifteen minutes, my calculations say that a row of grain sorghum 200 feet long can be expected to yield at least 1 bushel of grain. And that might be the right amount for your first venture into grain sorghum.

Storing the seed (threshed or in the bundles) poses no insect problems in the North, at least not in my experience. But in the South (or, I suppose, anywhere grain sorghum has been stored over a number of years), both the rice weevil and the Angoumois grain moth (no, I do not know how to pronounce that one either) can be problems, so say the experts. Like the bean weevil and some wheat weevils, these two characters can infest the grain before it leaves the field, which means that all those directions about storing in insect-proof cans are out the window. The insects are in the grain already. (For ways to solve the dilemma, see chapter 3, where I talk about grain storage.)

Feeding grain sorghum on garden farms that support only a small number of animals is easy. Just toss a few unthreshed seed clusters to the animals each day. With chickens, for example, serve up one seed cluster per six hens per day. Most animal nutritionists do not recommend feeding *only* grain sorghum unless you feed supplemental plant protein like soybean meal. But if your hens can graze outdoors, no problem. They will do their own supplementation with bugs and worms.

University of Georgia tests show that cattle on grain sorghum alone didn't do as well as on corn because the crude proteins in the sorghum aren't utilized well enough by the animals. When soybean meal was added, the sorghum feeding resulted in admirable gains of about 3 pounds per day, as good as from corn. Non-plant protein like urea won't produce the same good gains, report the scientists.

Sorghum Flour

For flour for your own use, thresh out a couple of measuring cupfuls when you need them. I take one seed cluster at a time, hold it over a bucket between the palms of my hands and rub back and forth vigorously, as if I were rolling out a ball of clay. Wear gloves. The seed will thresh out fairly easily, and you will get your two cupfuls quickly. You'll also get some hulls and stem bits, which you can winnow out in front of a fan (see chapter 3). You won't get your grain completely clean, but no matter. That wee bit of fiber from hulls and stems won't hurt in the flour—it might even do you good.

You can grind grain sorghum in the blender, but a regular kitchen mill is better. We mix it two-thirds sorghum flour and one-third wheat flour because, to our tastes, sorghum flour alone is too heavy, especially for bread. You be your own judge. For cookies you can use the sorghum flour alone with good results.

Today, grain sorghum recipes, like all whole-grain recipes, are easy enough to find. Arrowhead Mills, the well-known Texas organic food grower and distributor, has recipes to go with the sorghum grain and flour they sell. Check the Internet and prepare to be overwhelmed.

Jerry Logsdon feeding sorghum seedheads to chickens.

Sorghum Pancakes

1½ cups of sorghum flour (We use either grain sorghum or sweet sorghum seeds, which we grind in our blender.)
2 tablespoons brown sugar
1 teaspoon salt
3 teaspoons baking powder
2 egg yolks
3 tablespoons melted butter
1¼ cups milk
2 egg whites, beaten

Sift and mix dry ingredients well. Beat egg yolks, then add the melted butter and milk. Fold in the beaten egg whites and bake on a hot greased griddle, turning once when the pancakes are all bubbly on top. Serve with maple syrup if you have it, or any other syrup or fruit jam you like.

Sweet Sorghum

Sweet sorghum looks just about like grain sorghum at maturity except that it is about three times taller. It is grown for the juice in its stalk, from which is made sorghum syrup and sorghum molasses. I call it the four-in-one plant because it actually produces four foods. The seed clusters on top can be used just like grain sorghum, for flour and for chicken feed. The leaves, stripped off before the stalks are pressed, can be fed to livestock. Then there's the distinctively flavored syrup. And the crushed stalks left over make good mulch for soil microorganisms to feed on.

In fact, now that scientists in the USDA's Agricultural Research Service have discovered how to remove starch from sweet sorghum syrup, sweet sorghum can be processed into sugar in a conventional sugar-processing factory. Perhaps someday in the North where sugar cane doesn't grow, this process might make every homestead free of dependence on stores for sugar. So far this has not happened because sugar from sugar cane is heavily subsidized and more efficient to process.

An acre of sorghum will produce about 400 gallons of syrup,

which at a retail price of $8 a gallon is a nice sideline income for a small homesteader and a few weeks of work. Traditionally, that's how most of the sorghum syrup has been produced, and still is. A small farmer can handle an acre or two mostly with hand work and then haul the stalks to the neighborhood sorghum mill, which processes the syrup for a fee.

Again I hate to talk about varieties because they change constantly. But with sweet sorghum that is not so much the truth. The old varieties are still the ones mostly used, like 'Wiley', 'Brandes', 'Honey', and 'Sart'. In the North, you have to go to a shorter-season variety. 'Dale' and 'Tracy' are recommended varieties for central to northern regions. Farther north, I've grown an old variety, 'Sugar Drip', from R. H. Shumway Seeds. That's the only variety I've found easily available here.

The cultivation of sweet sorghum is much like that of corn. One exception is that sorghum doesn't require as much extra fertilization as corn to get adequate yields. That goes for grain sorghum too. Research in Texas has shown that sorghum fertilized with manure at the rate of 10 tons per acre for five years yielded just as well or better than sorghum on which chemical fertilizers were applied for maximum yield. University of Mississippi tests suggest that 40 pounds of actual nitrogen, plus 20 pounds of actual potassium, is sufficient for sweet sorghum, and a 5-ton application of manure should provide that. For the organic grower, a green-manure crop plowed under ahead of the sorghum crop and a couple of tons of finely ground rock phosphate every three or four years will provide the phosphorus and other nutrients the sorghum needs. Just remember that rock phosphate only releases nutrients very slowly. Chemical superphosphate is much more potent—one of the steroids of the agronomy sports world.

Cottonseed meal and tankage, both quite high in nitrogen and phosphorus, were once the favorite fertilizers of sweet sorghum growers. But competition for these products from the feed industry and the advent of cheaper chemicals pushed the price of these materials too high except for very small plantings. But some small growers still use these organic fertilizers, insisting that organically fertilized sorghum makes a better-tasting syrup.

You can plant sweet sorghum in rows or in hills. In hills, plant four seeds per hill with hills spaced 24 inches apart. In a continuous row, try to keep plants at least 6 inches apart. Any thicker, and the stalks might be weak and blow over.

Rows are generally spaced 42 inches apart. That was the traditional space needed by a horse or mule to get down between the rows pulling a cultivator. Some growers like to plant in a slightly raised ridge that they make with a kind of plow called a middle-buster, ahead of planting. Right before planting, they scrape the top off the ridge, which effectively kills the weeds germinating there. The raised ridge warms up a little faster in springtime, too, to enhance germination of the sorghum. Sweet sorghum, like grain sorghum, will rot rather than germinate in cold wet soil. Soil temperature at planting depth (2 inches) should be nearly 70°F on a sunny day. You need at least 100 days to mature sweet sorghum in the North, so the sooner you can get it growing, the better. The best sorghums take longer than 100 days to mature, which is why the best syrup comes from the South.

Some of the same insects that attack corn can also harm both grain sorghum and sweet sorghum, though the damage will rarely be serious: lesser cornstalk borer, corn leaf aphid, fall armyworm, corn earworm, and wireworms. In addition, the sugarcane borer will tunnel into sweet sorghum in the Deep South. The sorghum midge occurs over most of the Gulf Coast and the Southeast Atlantic states, and feeds on the heads of sweet sorghum, grain sorghum, broomcorn, and Sudan grass. But its depredations don't seem to hurt syrup production much. Where you are saving sorghum seed for seed or for food, however, and the midge is on the prowl, you can tie paper bags over the heads of plants during the blooming season. But, says the USDA, remove the bags shortly after the blooming period because they provide conditions favorable for the corn earworm and the corn leaf aphid, which also attack the seed heads.

You can judge the proper harvesting time for the juice by the condition of the seed. When the seeds are no longer milky, but still in the doughy stage, it's time to cut. When your thumbnail will no longer dent the hardening seed easily, the ideal harvesting time for syrup has passed.

First, strip the leaves from the stalks, and feed them to cows, rabbits or goats. Then cut the stalks off close to the ground, using a corn knife as I described when cutting a bundle of cornstalks.

When I can hold no more stalks in my left arm, I drop them neatly in a bundle and proceed to cut another bundle. There's no real need to tie the bundles, though that makes them easier to handle.

With the stalks in neat bundles, I can sever the seed clusters from the stalks with one stroke of the corn knife. Well, sometimes it takes two strokes. I usually lay the stalks on the tailgate of the pickup when it is down, with the clusters hanging over the edge so I can get a good whack down through the stems. Then I tie the clusters together and hang them in the barn for further drying.

The stalks are run through the sorghum press, which squeezes out the juice the same way the old clothes wringers squeeze water out of clothing. The juice that makes the syrup is held in the soft inner stalk and is easily pressed out once the tough outer stalk is cracked open by crushing. The rollers are powered by reduction gears running off a gasoline or electric motor, or by a horse or mule walking round and round on the end of a long sweep.

Sorghum mills press the "cane" and boil down the juice for a fee. There are still quite a few of these mills in the South, but here in the North, before we found one in a nearby Amish community, I had to improvise another way to squeeze out the juice. My method was not very practical, but maybe it will give you a better idea. On my first try, I chopped up the stalks with the corn knife into little pieces on a makeshift chopping board, sort of the way you would dice carrots. Then I squeezed the choppings in a cider press. It worked, but the chopping was slow and arduous. A wood chipper might work better, but I didn't have one.

The liquid from the pressing is a cloudy green in color. When boiled, heat coagulates the nonsugar materials in the juice, and they float to the top. These "skimmings" as they are called, have to be dipped off or the syrup will taste too bitter.

We heated ours in a pot on the stove and skimmed off the green stuff with a ladle. The syrup reaches the proper density at a temperature of 226° to 230°F. If you are using a syrup hydrometer, syrup of good density should give a reading of 35° to 36° Baume on the hydrometer scale. Before filling small containers, let the syrup cool to at least 190°F, 180°F for larger containers.

Traditional sorghum making is a far more picturesque and romantic procedure than our stovetop venture. As the horse or mule walks round his endless circle on the sweep, the mill rollers crack the stalks and squeeze out the juice, which runs through several strainers into a barrel. From the barrel, a hose or pipe gravity-feeds the juice into the steaming evaporator pans, which rest over a long, low fireplace with a high smoke pipe at one end. The fires underneath are fed with firewood and kept at as constant and even a flame as possible.

The evaporator pan usually is separated into three or four compartments by baffles, which allow the sap to flow from one section to another very slowly. As the syrup flows from one end of the long tray to the other around the baffles, the "cook" skims off the green scum. When everything is moving as it should be, the syrup is ready by the time it reaches the last compartment. If you lift the ladle out of it, the syrup should "string," showing that it has the proper density. It is ready then to run off into the waiting container, to be cooled and bottled.

One secret of sorghum making is to cool it as quickly as possible. That's why, in the past, homesteaders liked to locate their sorghum mills near a good cold spring. They'd dunk the containers in cold water.

Weather can be a factor too. A clear, crisp autumn day is ideal. Rainy weather makes bitter molasses, according to folklore. Maybe that explains why one batch of our sorghum turned out delicious and the other had a taste like licking an old piece of metal siding.

The seed clusters cut from the stalks and dried in the barn can be fed to chickens as described above for grain sorghum. The seed can likewise be threshed for flour. I think it tastes about the same as grain sorghum. Sorghum pancakes and sorghum molasses—now there's real poetry for homestead gourmets.

As to which variety of sweet sorghum makes the best flour or the most seed for flour, I don't know. 'Sugar Drip' is the old tried-and-true variety for syrup, and I have not heard of anyone making flour from any other kind of sweet sorghum. From the appearance of the seed clusters of other varieties, I'd say they would all serve the purpose except maybe 'Honey', which doesn't seem to have as much seed in its head.

Broomcorn

Broomcorn reached its heyday in the early 1900s, then declined as cheap synthetic materials took much of the market. But traditional brooms seem to be making something of a comeback, as decoration as well as for sweeping up dirt. (The new demand may be coming from our politicians, who always need prodigious numbers of new brooms to sweep clean the floors of preceding administrations.) Maybe the new interest in herbs has something to do with it. Botanical.org on the Internet says that a decoction of broomcorn seeds has medicinal value as a diuretic. Put 2 pounds of seeds in a quart of water, boil down to a pint, and you have a potion for urinary problems, say the directions.

Real broomstraw makes a better broom than plastic straws, but that depends on who is doing the sweeping. My grandsons could not do a good job of sweeping with the best broom ever made. But in these days not so many people care about the finer skills of housekeeping. Why pay $5 more for a good broom, when a poor one will do? The answer is that those of us who find all kinds of uses for brooms in the barn where an electric sweeper just isn't appropriate still appreciate a real broom.

I happen to have a catalog from the 1920s advertising all the tools of broom making. The number and variety are remarkable until one realizes how big the market once was. Every household, shop, and factory had to have brooms. No telling how many people made their living in the industry. If the energy crunch gets worse, natural broomstraw and the efficiencies to be gained from local economies just might make broomcorn a big industry again. You can get a thousand brooms from an acre of broomcorn. Takes a lot of work, but so do strawberries. It's something to think about, anyway. Need I have to say that today most of the broomcorn for brooms comes from China?

We have grown a little broomcorn, mostly for fun. The chickens don't like the seeds as well as those of other sorghums, but wild birds do. One year we tried our hand at making brooms from our crop, the old "round" brooms, the kind you see witches riding around at Halloween time, or the kind of broom now called a fireplace broom. They make nice decorations, hanging from the mantels in fine houses where they never sweep up anything. We had no special broom-making machinery but followed *Reader's Digest*'s 1981 book, *Back to Basics*, which has illustrated, step-by-step instructions on how to do it. We gave the brooms as Christmas presents in place of store-bought items that would no doubt have cost us $20 or so each. Or $50 today.

For handles, we selected young straight hickory or ash saplings from the woods, shaved them down, and pointed them at one end. We combed the seeds out of the broomcorn tassels with a wire brush, leaving about four to 6 inches of stem on each tassel. Then we soaked the tassels in boiling water and, while still wet, tied them very tightly around the end of the handle. A couple of nails driven through the handle will help anchor the tassels while you tie them. To make a tight tie, the trick is to hang a bit of rope with a loop in the bottom from something solid on the ceiling. The loop should hang about a foot off the floor. Twist the rope once around the broom stems to be tied, put your foot through the loop, and step downwards. The rope will constrict around the broom stems and broom handle very tightly. Then tie them. We tied two courses of tassels around the handle, one a little lower than the other, and then trimmed the straws even. As the wet stems dry, they shrink around the handle.

Broomcorn is not difficult to grow. Being a member of the sorghum family and related to corn, it will grow where corn will, and by the same cultural methods. But broomcorn is more at home in a hot, dry climate than a humid one. Follow the directions given for sweet sorghum above: about one plant every 6 inches in rows 30 to 40 inches wide. You will almost always get the seed planted too thickly because it is small, so be sure to thin after the plants are up. Cultivate as often as necessary to control weeds.

For best quality, the brushy heads should be cut when the plants are in late bloom stage (before the seeds have developed fully). Cut off the top brush, leaving 6 inches of stalk below the brush. Dry these brushes for three weeks before using them for broom making. If you let the plants mature before cutting the heads, the straws turn yellow and many seeds form. The brushy seed clusters make an attractive dried-flower arrangement.

No machine was ever contrived to harvest broomcorn heads satisfactorily. That's why I think the crop still has potential for homesteaders. No one else is fool enough to want to work that

Carol Logsdon stripping leaves off sorghum stalks.

hard. Workers go through the standing stalks and break over the heads. Later the heads are cut or jerked off, the seed threshed out, the broomstraw cured for several weeks, then baled and shipped to factories. In older times, the seed was scraped out of the heads instead of threshed, and I have found it necessary to scrape out some seeds in my little broom-making experiments too.

The broomstraw used to be classified “insides and covers,” “self-working,” and “hurl.” Insides and covers (short straws) were used for underwork in the broom, though a *good* broom has almost all long straws. Hurl was, and is, the long straw. Self-working was broomcorn that contained no waste. Self-working was what everyone hoped to get, but seldom did.

If you want more information and a source of seed, contact Seed Saver’s Exchange (3076 N. Winn Road, Decorah, Iowa, 52101). You should know about Seed Savers anyway because it is an excellent place to buy all kinds of grains and heirloom vegetables. Or there may very well be a broom maker in your community since new ones are constantly taking a try at it. You could learn something from him or her, and perhaps become part of an interesting cottage industry. Let’s see, 1,200 brooms per acre at $10 per broom is . . .

CHAPTER FIVE

Oats

The High-Protein Cereal Grain

The Scots did not think Samuel Johnson was so cute when he defined oats in his dictionary as "a grain which in England is generally given to horses but in Scotland supports the people." A snide remark Johnson may have intended, but he was actually paying the Scots a high compliment. They had not only discovered a food very high in nutrition and very low in cost of production, but one that grew well in their climate. The first commandment of the wise garden farmer is to learn how to make good food out of what grows well in one's own climate.

Of all the cereal grains, oats ranks highest in protein and runs neck and neck with wheat as the all-around most nutritive cereal grain. And as everyone knows now, oats are considered to be particularly helpful in lowering cholesterol. The 1950 USDA handbook on grains rates oats at 14.5 percent protein, while whole wheat runs second with 13.4 percent. These figures are somewhat outdated now, especially in regard to oats. The average of 287 varieties selected from the World Oat Collection recently averaged 17 percent in protein content. More significantly, new varieties are constantly raising the protein ante to as high as 22 percent on a dry basis. That could make oats almost competitive with soybeans in terms of protein (soybeans contain about 35 percent protein but yield less per acre than oats).

Most plant scientists express belief in a bright future for oats as human food. Part of their reasoning is based on the character as well as the quality of oat protein. It has a bland taste, is soluble under acidic conditions, is stable in emulsions with water and fat, and holds moisture, thus making it an ideal protein to supplement other foods. At the USDA laboratory in Peoria, Illinois, researchers are using oat protein to make nutritious refreshment beverages, meat extenders, and high-protein baked goods.

Oats

Oats also outscore other cereal grains in thiamine, calcium, iron, and some say, phosphorus, though the USDA tables from 1950 in table 4 give whole wheat a slight edge in that department. Whatever, the table makes a good comparison chart of the benefits of all grains.

If you keep horses, sheep, or rabbits, it will pay you to grow oats. Even a small patch in the garden will save money on your rabbit-feed bill. You don't even have to thresh the grain out, just feed it, stalks and all, to these animals or other kinds of livestock. Recent experiments by Ohio State in the southern part of the state indicate that oats, sown in late summer, can provide high-quality winter grazing for cattle and sheep.

Oats are good livestock feed when ground or rolled and mixed with ground corn. That's the standard dairy-cow ration on most U.S. farms. Whole oats are excellent food for poultry, too; the hulls help prevent cannibalism. Chickens will eat more oats if they are rolled and mixed with milled corn and wheat. Given a choice between wheat and oats in their whole-grain scratch feed, they will invariably eat up all the wheat first. The oat hull that covers the groat so tightly on each grain may be good for them, but I suspect—as in the case with humans—the taste is not.

Oats make fairly good hay as well, though it takes a little longer to dry after mowing. Cut them when the grain is in the early milk stage, just beginning to fill out and the plant is almost entirely green yet. Yields of 8 tons per acre are possible or 600 pounds of plant protein per acre—not as good as alfalfa hay, but remember that you have to wait a whole year to harvest alfalfa after you plant it. You wait only two months for hay from a planting of oats. Be aware also that oat straw traditionally has been fed to cattle, especially oat straw stored in a stack. Such straw often still has a streak of green in the stalk, and hungry cows will eat it readily, especially when they are being fed a lot (too much) of corn.

Types and Varieties of Oats

Common white oats, *Avena sativa*, are by far the most widely grown species. They are planted in spring for midsummer

TABLE 4:
NUTRITIVE VALUE OF 1 CUP OF UNCOOKED GRAIN

	Recommended Daily Allowance (women aged 31–50)	Recommended Daily Allowance (men aged 31–50)	Barley	Buckwheat	
Calories			651	583	
Protein (g)	46	56	22.96	22.52	
Fat (g)	20–35	20–35	4.23	5.78	
Carbohydrate (g)	130	130	135.20	121.55	
Fiber (g)	25	38	31.8	17.0	
Calcium (mg)	1,000	1,000	61	31	
Iron (mg)	18	8	6.62	3.74	
Magnesium (mg)	320	420	245	393	
Phosphorus (mg)	700	700	486	590	
Potassium (mg)	4,700	4,700	832	782	
Zinc (mg)	8	11	5.10	4.08	
Copper (mg)	0.9	0.9	0.916	1.870	
Manganese (mg)	1.8	2.3	3.575	2.210	
Thiamin (mg)	1.1	1.2	1.189	0.172	
Riboflavin (mg)	1.1	1.3	0.524	0.723	
Niacin (mg)	14	16	8.471	11.934	

harvest. Red oats, *Avena byzantina*, are the southern and south-central type, sown in the fall where winters are mild, for harvest the following summer. Hull-less or naked oats, *A. nuda*, are a rarely grown third type. In addition, there are the kinds of oats that young men sow before settling down to responsible adulthood. It's a matter of opinion whether these oats are good or bad for society.

The distinction between white and red oats is often hazy except for the difference in planting time. White oats are more yellow than white, and red oats are often grayish in color. Many of the common white-oat varieties have red-oat parentage somewhere in their background. Clinton, one of the more successful white-oat varieties over the years, is actually a cross between white and red. 'Cherokee' and 'Andrew' are red varieties marketed as white, and 'Missouri O-205' and 'Kanota' are quite similar varieties marketed as red. The obvious deduction the reader should make is that vari-

Corn-meal (yellow)	Oats	Brown rice	White rice	Rye flour	Whole wheat flour	White flour	Wild rice
442	607	688	716	361	407	455	571
9.91	26.35	14.25	13.00	9.58	16.44	12.91	23.57
4.38	10.76	5.09	1.04	1.81	2.24	1.23	1.73
93.81	103.38	144.72	158.30	79.04	87.08	95.39	119.84
8.9	16.5	6.5		14.9	14.6	3.4	9.9
7	84	63	6	24	41	19	34
4.21	7.36	3.42	1.60	2.16	4.66	1.46	3.14
155	276	272	46	76	166	28	283
294	816	502	190	211	415	135	693
350	669	509	152	347	486	134	683
2.22	6.19	3.84	2.20	2.03	3.52	0.88	9.54
0.235	0.977	0.526	0.420	0.293	0.458	0.180	0.838
0.608	7.669	7.112	2.074	5.569	4.559	0.853	2.126
0.47	1.19	0.785	0.14	0.293	0.536	0.15	0.184
0.245	0.217	0.082	0.096	0.116	0.258	0.05	0.419
4.431	1.499	8.185	3.2	1.762	7.638	1.562	10.773

Sources: For nutritional content, USDA Nutrient Data Laboratory online database at www.nal.usda.gov/fnic/foodcomp/search/. For recommended daily allowances, National Academy of Sciences, Institute of Medicine, Food and Nutrition Board, "Dietary Reference Intakes (DRIs): Recommended Intakes for Individuals, Vitamins" at www.iom.edu/Object.File/Master/21/372/0.pdf, "Dietary Reference Intakes: Macronutrients" at www.iom.edu/Object.File/Master/7/300/Webtablemacro.pdf, "Dietary Reference Intakes : Electrolytes and Water" at www.iom.edu/Object.File/Master/20/004/0.pdf, and "Dietary Reference Intakes: Elements"" at www.iom.edu/Object.File/Master/7/294/0.pdf.

ety in oats (and maybe most grains) is not that important, at least for the garden farmer.

Many varieties of both white and red oats are available, but it's not necessary for you to know them by name. That's true of all cereal grains. The life of any one variety is apt to be quite short, limited to the five to ten years it takes for a disease pathogen to adjust to that variety's inbred resistance. That's why plant breeders have to develop new varieties with new resistances to disease constantly, and why I am loath to name names. Some have better milling quality, as millers would say, but I don't think that is important for small-grain raisers. I've read that Quaker Oats goes right to the marketplace for its supply, selecting high-quality grain of whatever variety.

Oats respond to ample rains more than other cereal crops to make a good yield. They like fertile soil too (what doesn't?) but will perform satisfactorily without high additions of fertilizer. In fact, you can easily get too much nitrogen in the ground for oats, especially in a garden situation where you've made the soil very rich. If a previous oat crop possessed a dark green color and much of it fell over on the ground before harvest, I wouldn't add any fertilizer. Even on a commercial basis, if oats seldom lodge and maintain a healthy green color, add no more than 30 pounds of nitrogen and 10 pounds of phosphorus and potassium per acre. If oats are short and light green in color, ripening to a flat tan rather than a solid yellow, the soil needs about 60 pounds of nitrogen, 15 pounds of phosphorus, and 20 to 30 pounds of potassium. So sayeth the experts.

Such low amounts of fertilizer can be supplied organically at reasonable cost even on large fields. Legumes will fix that much nitrogen in the soil naturally. So a green-manure crop with some animal manure added, along with 2 tons of rock phosphate applied per acre every three years and about 2 tons of lime every five to ten years, should result in yields of 60 to 70 bushels per acre and more on good ground.

Fertilizer balance is the key. After applications of either organic fertilizer or inorganic chemicals, soil tests may show a field contains 150 pounds of available nitrogen, but only 30 pounds of potassium. That could spell trouble for a cereal crop like oats. It would grow big and heavy and fall over so flat on its strawy back that it can't be picked up even by the most sophisticated combine header. If you have 150 pounds of nitrogen available, you ought to have at least 60 pounds of potassium available to give the stalks enough strength to match the heavy growth the nitrogen will cause. The garden farmer should think the other way. If you have 30 pounds of potassium, 60 pounds of nitrogen is enough. After all, you are not trying to pay your way to Bermuda on your oat profits.

Insects and Diseases

Oats have fewer enemies than corn or wheat: no Hessian fly, no chinch bug. Sometimes greenbugs, a type of aphid, will attack oats, but in many years a small wasplike insect, *Adbidius testaceipes,* keeps them in check.

Of the diseases that attack oats, crown rust is probably the most important, especially in southern and central areas. There's no cure, just preventive maintenance. Use resistant varieties and cut out any buckthorn bushes near fields or gardens. The disease uses buckthorn as a host plant during part of its life cycle. That's what the books say anyway. But at least one kind of buckthorn, 'Cascara', is renowned in folk medicine as an effective laxative. Perhaps the backyard oats grower with chronic constipation might decide to risk it.

Septoria leaf blight is another fungal disease that recurs in cereal crops, including oats. Again, crop rotation and resistant varieties are your best defense.

Quality is as important in grains as it is in fruits and vegetables. The grower must strive to keep them clean of insects, mold, and any other foreign matter. But just as important, grains in commerce must meet certain weight standards to qualify as good, better, or best, and this is especially true of oats. A bushel of oats is supposed to weigh around 38 pounds, but only plump, healthy, well-grown oats will actually weigh that much. Test weights of 36 and 37 pounds per bushel are considered very good, but when test weight falls below 30 pounds, you are handling many shriveled grains that contain little food value and low germination potential. Weight of a grain can also tell you something about its moisture content. The drier the grain, the more it will weigh in any given volume, everything else being equal. That may surprise you, but the moisture that swells the grains so that fewer of them fit into a bushel weighs less than the grains themselves. Removing the moisture allows more grains per bushel.

Oats Culture

Whether you plant in fall or spring, raising oats proceeds in similar fashion. Since my experience has been only with spring oats, I'll describe that process. Southern growers can adjust what I say to their own fall-planting conditions, or proceed in a manner similar to what I described for planting wheat in the fall. For spring oats, the earlier you can plant the better. I have planted oats in Minnesota when there were still snowdrifts melting in the woods. Oats like cool weather and can get along just as well with cloudy weather as with constant sun. That's why they are adapted so well to a place like Scotland. In the North and East in the United States, oats are a good crop to grow wherever potatoes grow well. The two seem to like a similar environment.

"Farmers," says an old adage, "mud in the oats and dust in the wheat" to get good crops of both, referring to spring oats and fall wheat. The saying is more accurately a description of what usually happens rather than good advice. When you plant oats in the spring, you are usually battling wet weather, and when you plant wheat in the fall, you are contending with dry weather, whether you like it or not. Be that as it may, whenever the soil dries enough in spring to be workable, plant your oats. The longer you wait, the poorer your subsequent crop is likely to be.

If you have gardened a long time, you have noticed, I'm sure, that almost every year there is a short period of dry weather in early spring when the ground does dry out enough to till. The temptation, which most of us give in to, is to plant some early vegetables. About half the time this planting amounts to very little because the ground is still too cold for good germination, and more cold weather is going to come anyway. So instead of planting vegetables at that time, plant a patch of oats and you'll be ahead on both counts.

The ground you intend to plant oats on will probably have been plowed or disked or rotary-tilled the preceding fall. Fall-worked ground dries out faster in the spring than spring-plowed, so you can get on it earlier. The finer the seedbed the better, but for oats you can be a little less finicky and get away with it. Invariably you are going to get more rain shortly, which will ensure that the soil settles over and around the oat grains well enough for good germination.

In the garden, you can do it all with a rotary tiller. Work the ground up, but not too finely, broadcast the seed by hand, scattering it as evenly as you can over the plot, then run over it lightly again with your tiller. This light second tillage covers the seeds adequately—at least partially. You don't have to cover every single grain. With good moisture, the grains on top sometimes sprout faster than the ones covered.

On a larger plot or field that has been disked and harrowed, plant either by broadcasting or with a drill, as described in chapter 3. The seeding rate for oats is 2½ bushels per acre. The drill puts the grain in the ground and covers it automatically. Set the drill to plant the seed not more than 2 inches deep, and 1 inch is better in early spring. If you broadcast the seed, cover it by going over the ground very lightly with a disk or a harrow, or better yet, a cultipacker. You will get some seed planted deeper than 2 inches, and some barely covered at all, but don't worry. Enough will germinate normally. Broadcasting won't make as good a stand as drilling. That's why most broadcasters will plant at a 3-bushel-per-acre rate, a little higher than the norm.

Weeds will be a problem in larger oats fields unless you follow a good year-round, year-in-and-year-out program of weed prevention. Once the oats are planted, you can't get into them to cultivate, though on a small patch you can walk through and hoe out some weeds. If you are not farming organically, you can spray with a broad-leaved herbicide, which if used correctly won't harm grasses like oats. According to science, oats produces its own natural herbicide, but I have my doubts as to its effectiveness, because I have seen some mighty weedy oat fields.

Or you can plant, in the garden, in rows wide enough apart to get the cultivator through. In this case some weeds will grow in the row, but you can take out enough of them to keep the oats growing fine. If the field was weedy last year, you can be sure your oats will be weedy too. The weeds won't necessarily "take the crop" and might not even hurt your yield much, but they make harvesting more difficult and increase the problem of getting weed seed and weed chaff out of the grain.

You can harvest your oats as grain, as hay, or even as silage. For harvesting with a grain combine, wait until the crop is dead ripe. Or cut it, rake it as you would hay, and allow it to ripen in the windrow. The windrows are harvested with a special pickup attachment on the combine. In the more northerly states, this latter method is still used because farmers believe the grain ripens too slowly and unevenly in the uncut stalk. By cutting and windrowing, they can often get the grain harvested with less risk, since, if left standing to ripen, the grain may be knocked flat on the ground by a hard storm. (Of course, if it rains too long on the windrowed oats, part of the crop can be lost too. Farming is always a gamble.)

I know a farmer who used to cut and bale his oats as he did hay, and then feed it by the bale to his animals. His livestock ate the oats and some of the straw. The rest of the straw became bedding. Since oats was the only small grain he grew, this practice saved him from buying a combine.

You can cut the oats when the grain is just beginning to harden, and the stalks have still a little green in them. Tie the stalks into bundles, as described with wheat in chapter 3, and place the bundles into shocks, where the oats can finish ripening and drying somewhat protected from rain. Then rank the bundles in a barn or even outside like a double stack of wood with the butts of the bundles to the outside and the heads inside to protect them from rain. Feed the oats by the bundle as needed.

You won't find this manner of harvesting, storing, and feeding oats advised anywhere else that I know of anymore. It's a method out of the past, which fortunately fits the homesteader of the future. I was pleased to learn that as late as 1963 (and without doubt still true somewhere) USDA officials observed small homestead plots of oats being harvested in the central states with grain cradles and (more often) with old-fashioned binders, the oats then fed by the bundle unthreshed to livestock. I'm not surprised, but I am glad I can now point to experience other than my own to substantiate what I know is a very economical and practical way to feed grain to animals. Not only do the animals "thresh" the grain as they eat it, but they clean up most of the straw too. Any old-timer will tell you that cows and horses like oat straw. They will sometimes eat it for roughage as well as they eat hay. It isn't as good for them as hay, we say, but who's to argue with a cow? At least they consume more total roughage that way, which is all to the good.

Mice will get into oats stored with the grain still in the straw. That's why you should feed these oats out through the winter as quickly as possible. Cats in the barn are a big help.

Oats for the Table

The oats you want for your own use you can thresh and winnow in the same way you would thresh wheat by hand. Thirteen-and-a-half bushels of good oats makes a barrel (180 pounds) of rolled oats, so you would hardly want more than a couple of bushels' worth, assuming you are using other grains too. If you are harvesting your oats with a combine, or more likely are having it harvested by a farmer who has a combine, you can, of course, just take the grain from the combine bin and winnow it cleaner if necessary.

Oats are more difficult to prepare for human food than wheat. The tight hull around the oat groat needs to be removed. There are grain elevators and mills that dehull oats, but finding them means doing research. Doing research gives me a chance to rant again about the problems of hunting for hard-to-find farming tools and services. Most readers already know what I am going to say. But let us entertain this scenario: you have been gardening for a long time and now realize that there is no good reason why you aren't growing grains as well as fruits and vegetables. You read what I am writing, but you have no other contact within farming circles, so you write to me for information or go on the Internet. That's okay, but what you really need to do is to get into the farming loop in your own local area. Everybody's talking about buying local food these days, and one of the best results of that effort is that it will force the urbanite interested in good food to get to know local farmers. I live in northern Ohio, and I am ill-suited to give you specific advice if you write to me from California. Get to know the farmers in your area. Go to farm-supply stores and ask around. Attend or join the small farm organizations in your area. Read magazines published specifically for small-scale farmers and garden farmers, such as *Farming Magazine: People, Land, and Community* or *Small Farmer's Journal.*

Just as an example, when I was frantically searching for information about hulling oats, an invitation came in the mail from the Northern Plains Sustainable Agriculture Society to a symposium and farm tour in North Dakota. Meaning no slur whatsoever, I am just not used to thinking of the Dakotas as being on the cutting (or hulling, in this case) edge of organic grain production. How wrong I was. The brochure contained a veritable gold mine of sources for organic grains and services. One of quite a few ads was for "Domestic Hulling." The email address given was buckwheat@iw.net, so I presume the reference was mainly for buckwheat hulling. But anyone who knows about hulling buckwheat surely knows about hulling oats. Another ad was from Organic Grain and Milling (www.ceresorganic.com) in St. Paul, Minnesota. These milling businesses serve mainly commercial growers, but nevertheless I bet that anyone who belongs to the Northern Plains Sustainable Agriculture Society knows where you can get oats hulled in their region.

The Internet bristles with requests for information on where small, kitchen-sized hullers can be obtained, which means that interest is high. I'm sure that when entrepreneurs realize this, they will rise to the occasion. One place online to check out is the Plant Sciences Department at the University of California, Davis. There are folks working there who are interested in making small threshers and hullers for a home or kitchen situation. By the time you read this, they may be able to guide you to fruitful sources.

Having said all this, if you want to go it alone, here are some hints on how to get that oat hull off yourself. Commercial processors have found that heating the grain for one-and-a-half hours at a temperature of 180°F makes the hulls brittle and easier to remove. The heating also dries the grain down to around 8 percent moisture from storage moisture of about 12 percent, which helps maintain quality during storage. You can certainly do small batches by roasting your own oats in your oven.

One older method of removing the hulls after roasting is to grind them lightly between two carborundum or emery-stone disks moving in opposite directions. The disks have to be set very precisely, so that the space between them is just small enough for the stones to tear and scrape loose the hulls of the oats without pulverizing the groats.

In commerce, centrifugal, impact-type hullers use a high-speed rotor to throw the oats against a rubber liner hard enough to knock the hull loose and blow it away. Another way is to pass the oats through extremely sharp, whirling steel blades. The hulls are then winnowed from the groats.

The groats are steamed and passed through steel rollers for flaking and dried for old-fashioned oatmeal. Nowadays, when no one has time to enjoy a leisurely breakfast, we have oat flakes that will cook in three minutes or less, so that we have time to get to the office early and brag for three minutes about how we got to work before anyone else. To cut the time of cooking oats, the processors "steel-cut" the oat groats. All that means is that the oats are partially ground. Each groat is cut approximately into three parts.

Your blender will cut up the groats to any size you want, but it cuts up the hulls too. All that fiber may be healthful, but not tasteful. You can run the blender briefly and sift out or winnow out some of the hulls to get something approximating good oat flour.

You can set the "concaves," as farmers call them, close enough to the whirling threshing cylinder on a grain combine so that the grains are not only knocked out of the stalks, but the groats out of the hulls. If roasted oats were run back through a combine adjusted that way, I believe dehulling would be nearly complete, but I've never tried such a trick. A more practical method would be to run the roasted grain through any kitchen mill and winnow or sift out the loosened hulls.

Fortunately, there is another way out of the dilemma. Earlier I mentioned a rarely grown oats, *Avens nuda,* or naked oats. This species does not have the tight hull around the groat. Although it is available from many northern farm-seed companies, it has not become mainstream because it doesn't yield as highly as other oats and because birds love it. Both of these disadvantages make naked oats a good possibility for small-scale growers not looking for top yields and with small enough patches to protect them from the birds. After first promising my wife that I would not stare lasciviously at the crop when it matured, I grew naked oats many years ago. I planted about two acres with visions of money dancing in my head. I had an old Allis-Chalmers All-Crop grain combine then, and I intended to plant the seed I got on maybe ten acres and sell that crop for seed to homesteaders and garden farmers, whom I was convinced would want to grow their own oats someday. The birds swarmed in when the groats were in the milky stage and ate most of it.

But I should not have been so discouraged. Oats should yield an average 60 bushels to the acre—sometimes the yield is 100 bushels per acre. So, on a garden farm, one-twelfth of an acre should easily mean 5 bushels of oat groats. That will be enough for quite a few breakfasts. One-twelfth of an acre would take a plot of ground about 60 by 60 feet, which would not be too difficult to protect with bird netting during the grain's milky stage.

A couple of other interesting tidbits about oats may be of interest to you. A common practice among strawberry growers used to be to grow oats in the strawberry patch for mulch. Instead of having to buy straw and transport it to the garden, gardeners broadcast-sowed the oats over the entire strawberry patch *in the early fall or late summer.* The grain would grow tall but would not have enough time to produce seed before frost killed it. Dead, the oat plants fell over and maintained a protective mulch over the berry plants.

On a more modern note, the University of Minnesota several years ago was experimenting with new ways to grow edible mushrooms. They reported that the mushrooms grew quite well in a "soil" composed almost entirely of oat grains.

Oat hulls, as a by-product of the oatmeal industry, are used to produce furfural, an important industrial solvent. Also, oat hulls have been used to polish the pistons of upscale cars like the Rolls-Royce. That's a nice detail you can use to impress your friends as you feed them a homemade oat cake.

Oat Recipes

Peanut Butter Sesame Balls

¾ cup peanut butter
½ cup honey
1 teaspoon pure vanilla extract
¾ cup nonfat dry milk
1 cup oatmeal
¼ cup toasted sesame seeds*
2 tablespoons boiling water
Chopped nuts or toasted sesame seeds for coating balls

- Preheat oven to 200°F.
- In a medium-sized bowl, combine peanut butter, honey, and vanilla extract; blend thoroughly.

- Mix nonfat dry milk and oatmeal together. Gradually add to the peanut butter-honey mixture, blending thoroughly, using your hands if necessary to mix as the dough begins to stiffen. Blend in the toasted sesame seeds.
- Add 2 tablespoons boiling water to mixture, blending well.
- Shape into 1-inch balls. Roll in finely chopped nuts or toasted sesame seeds. For variety, roll half the mixture in chopped nuts and the other half in toasted sesame seeds.

Yield: approximately 3 dozen balls

**Toast sesame seeds in oven for about 20 minutes or until lightly browned.*

Sesame Crisp Crackers

1 cup oat flour (oatmeal may be ground in electric blender)
¾ cup soy flour
¼ cup sesame seeds
¾ teaspoon salt
¼ cup oil
½ cup water

- Preheat oven to 350°F.
- Stir together flours, seeds, and salt. Add oil and blend well. Add water and mix to pie-dough consistency.
- Roll dough on greased baking sheet, to ⅛-inch thickness. Cut into squares or triangles and bake in oven until the crackers are crisp and golden brown, about 15 minutes.

Yield: 3 to 4 dozen crackers

Almond Crunch Cereal

3 cups oatmeal
1½ cups coconut, unsweetened
½ cup wheat germ (or soy grits, if preferred)
1 cup sunflower seeds
¼ cup sesame seeds
½ cup honey

¼ cup oil
½ cup cold water
1 cup slivered, blanched almonds
½ cup raisins (optional)

- Preheat oven to 250°F.
- In a large mixing bowl, combine oatmeal, coconut, wheat germ or soy grits, sunflower seeds, and sesame seeds. Toss ingredients together thoroughly.
- Combine honey and oil. Add to dry ingredients, stirring until well mixed. Add the cold water, a little at a time, mixing until crumbly.
- Pour mixture into a large, heavy, shallow baking pan that has been lightly brushed with oil. Spread mixture evenly to edges of pan.
- Place pan on middle rack of the oven and bake for 2 hours, stirring every 15 minutes. Add 1 cup slivered almonds and continue to bake for ½ hour longer, or until mixture is thoroughly dry and light brown in color. Cereal should feel crisp to the touch.
- Turn oven off and allow cereal to cool in oven. If raisins are to be added to cereal, do so at this point.
- Remove cereal from oven, cool and put in a lightly covered container. Store in a cool, dry place.
- Serve plain or with fresh fruit.

Yield: 5 to 6 cups

Traditional Irish Oatmeal Bread

8 teaspoons dry yeast
1 cup lukewarm water
1 tablespoon honey
¼ cup nonfat dry milk
1 cup water
½ cup oil
1½ teaspoons salt
4 tablespoons honey
2 eggs, well beaten

2 cups oatmeal
6½ cups whole wheat flour
1 cup currants
1 egg, slightly beaten
½ teaspoon water

- Dissolve yeast in 1 cup lukewarm water. Add 1 tablespoon honey.
- Combine nonfat dry milk and 1 cup water with wire whisk, and heat almost to scalding point. Add oil, salt, and honey. Cool to lukewarm.
- In a large mixing bowl, combine milk mixture, 2 well-beaten eggs, and yeast mixture. Mix in oatmeal and 6 cups of the whole wheat flour, 3 cups at a time, reserving ½ cup for the second kneading.
- Knead until smooth and elastic, for 5 minutes.
- Put into an oiled bowl. Cover with damp cloth and let rise in a warm place until double in bulk, approximately 1½ hours.
- Stir dough down and knead with remaining ½ cup whole wheat flour, gradually working in currants. Shape into three round loaves. Brush the top of loaves with beaten egg to which ½ teaspoon water has been added. Put loaves on oiled cookie sheets to rise.
- Let rise 1 hour in a draft-free spot. Meanwhile preheat oven to 375°F.
- Bake in oven for 25 minutes until golden brown. Remove from pan and cool before slicing.

Yield: 3 round loaves

CHAPTER SIX

Dry Beans

The "Poor Man's Meat"

I don't know who first called dry beans the poor man's meat, but today they are also the rich man's meat, as many people take advantage of their rich content of proteins as a way to eat less meat or no meat. Technically, edible dry beans like the soybean and the navy bean are legumes, not grains, as everyone knows, but they belong in this book because growing them is much like growing grains, and the two make great partners both in crop rotations and in our diets.

In fact the basis of natural, sustainable farming is the working partnership between grasses (grains) and legumes (beans and clovers). Grasses grow well in rotation with legumes because the legumes draw nitrogen from the air to invigorate the grasses. When the grasses (grains) use up that free nitrogen and weaken a little, the legumes or clovers come back strong and charge the soil with more nitrogen. Farmers who take advantage of this partnership to its fullest can avoid spending lots of dollars on expensive nitrogen fertilizer and save inestimable amounts of natural gas, which is used extensively to make that fertilizer. To me there is no greater gift from nature than the combination of bluegrass and white clover because both will grow free and spontaneously in good soil and can provide the major part of the diet of grazing animals. Pasture farmers know this, and are again producing meat and milk almost entirely on the strength of permanent or semipermanent stands of grasses and legumes.

By the same token, corn and soybeans, or wheat and lentils, to name two examples, can be beneficial partners in farming. At least as long ago as Virgil, who sang the praises of partnering grains and legumes in his *Georgics,* this truth of good farming has been recognized. I believe that the garden farmer would do better to heed the words of poets like Virgil (or Wendell Berry today) than the money-inspired science of modern farming.

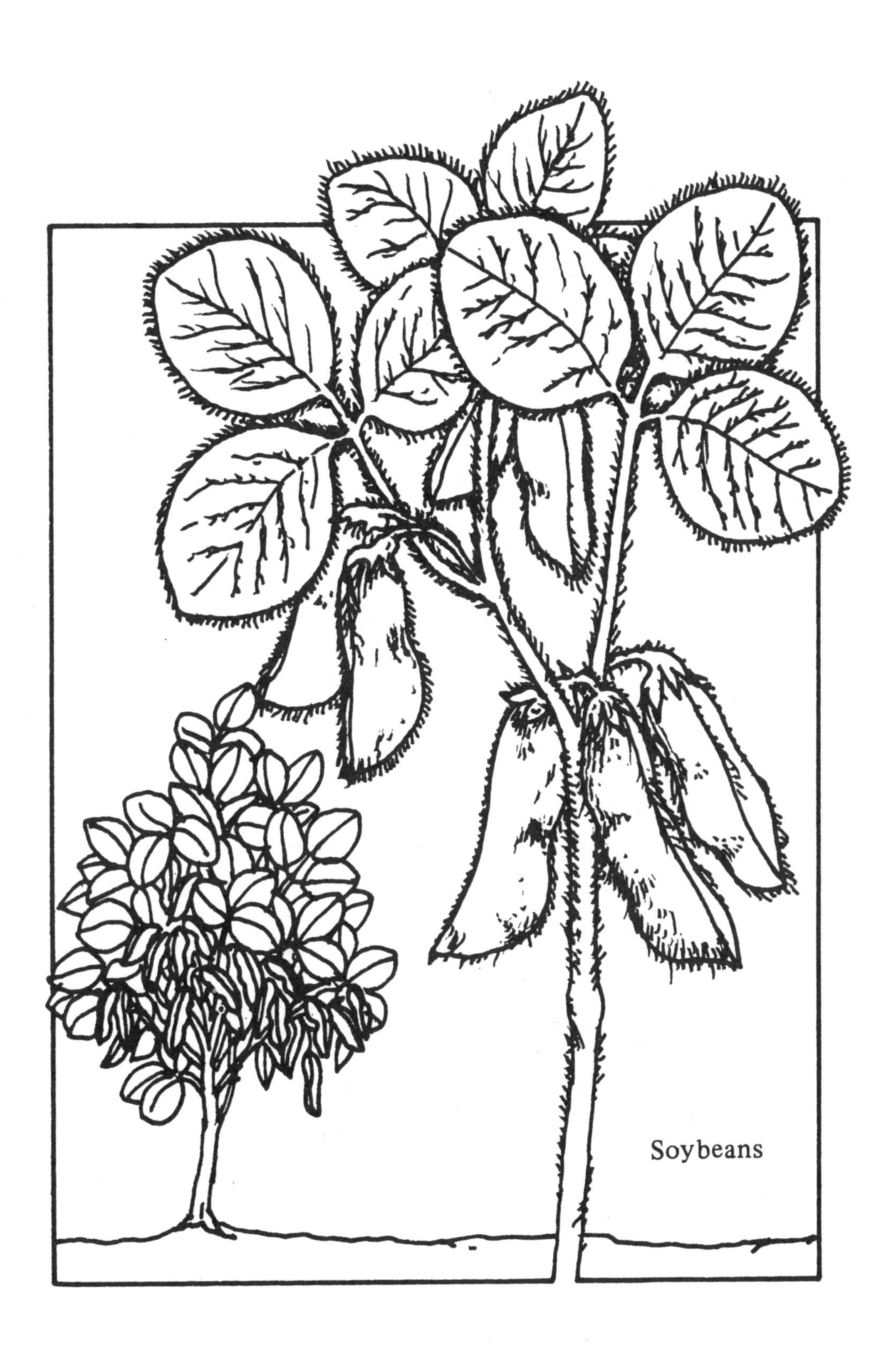

Soybeans

The sheer number of the edible dried beans shows their importance to the human diet around the world: soy, pinto, lentil, black-eyed, black, brown, kidney, white, sprouting, runner, tepary (I never heard of this kind until I saw it mentioned in a Seeds of Change catalog recently), garbanzo, chickpea, field pea—just to name a few. If I couldn't justify putting dry beans in a grain book any other way, then I would simply say that baked beans are one of the heavenly dishes of civilization and don't need any justification. If the carbon footprint folks want to worry that the subsequent gas from eating them is increasing global warming, I say when they quit exhaling carbon dioxide, I'll quit eating baked beans.

The soybean is the number one "cash grain" crop in America, so it gets the most attention in farming circles. I think soybeans can make wonderful food for humans, as Asian civilizations have shown for centuries, but I wonder, all things considered, if we should feed them to our farm animals when oats might be cheaper and generate a more sustainable kind of farming. Before the soybean came to America—"before farmers went crazy," as my father liked to say—American farmers grew corn, oats, wheat, and hay or pasture crops, in that order of rotation. Now much of the Corn Belt is in an endless rotation of corn and soybeans, which amounts to almost a monoculture. Fields in soybeans erode worse than fields in oats, and thus, with so many millions of acres in vast fields of soybeans, erosion problems are more severe. Soybeans do have more protein than oats, so they are a good food, properly prepared. On the other hand, because of the large quantities needed for farm animals, oats might be a better choice because they produce more grain per acre than soybeans and can be fed to animals without cooking. But, of course, oats do not put nitrogen in the soil the way soybeans do. So I guess it's a draw.

Raw soybeans should not be fed to animals or eaten by humans. I learned that the hard way. When I was a child, I tried to eat more soybeans right out of the bin than our hired man, who tried to make me believe he liked them raw. To this day that taste sickens me. Raw beans, especially soybeans, contain enzyme inhibitors and need to be cooked to get rid of them. Some nutritionists say that soybeans need to be fermented like the Japanese do for miso and tempeh rather than cooked, or in addition to being cooked. I am not going to get into that argument. Soybeans for animal feed are roasted or cooked in some way before they are crushed into soybean meal, which is then fed as a protein supplement along with grains. As far as I know, not much attention has been given to the cost of this roasting or cooking versus the much lower cost of feeding raw oats and good hay or pasture to animals for their protein. The small farmer can still avail himself of custom bean roasters who will come to the farm to do this operation, although, more often now, farmers or custom feed services have extruders to do the job more efficiently. An extruder heats the beans by friction and therefore cooks them as it crushes and grinds them into meal. Today commercial farmers mostly buy soybean meal from commercial suppliers and mix it with the grain rations for their animals. Fermenting the beans, or soaking and cooking them as for human food, would be very laborious for more than a very few animals. I think legume hay and legume pastures, with perhaps a little oats as a supplement to corn, is a better livestock feed than soybeans, and the manure won't stink as badly as manure from soybean meal.

With the exception of the delicious black-eyed pea or crowder pea of the South, most beans for baking are grown in the North, where they seem to do better. But evidence suggests that most of these beans can be grown equally well north, south, east, or west. Experiments at the University of Arkansas Experiment Station indicate that many dry bean varieties can be commercially successful in the northwestern part of that state at least. That means that they'd do all right in a garden even in the southern part of the state, and most likely on down to the Gulf Coast.

The reverse is true too. I've had good luck here in the North growing southern black-eyed peas, as well as northern white beans. They seem to produce with no extra care as to soil fertility. The crucial part is harvesting. I like to let the beans dry in the pod on the stalk completely, if possible. But if the weather is wet in September the beans are very apt to mold in the pod. The black-eyed peas seem to be worse than others in this respect.

Growing Dry Beans

Beans can be drilled carefully into finely worked seedbeds, no-tilled directly into mostly undisturbed soil following cotton or corn, broadcast by hand and disked roughly into the soil, even dropped on top of the ground from airplanes into standing wheat with no tillage at all. In all four situations the beans sprout and grow more or less successfully, although broadcasting them from airplanes is risky. The point is that if you make a mistake with soybeans or any dry bean, just say you did it on purpose. The beans will likely grow as long as ample rain comes. They will even withstand drought better than corn once they sprout and start growing. But plant them when the soil has warmed up well. (Some commercial grain farmers say you can get away with planting them in cold ground in April, but one often sees such farmers replanting their soybeans after the first planting doesn't come up very well.) Where the best corn planting date is May 5, figure May 20 for beans. And then you can go on planting beans until the middle of June or even later and the crop will still reach maturity before frost—at least below the 40th parallel. Being able to double-crop beans after barley or wheat can be advantageous not only for commercial farmers but also for garden farmers looking to use limited space to the fullest. But planted in July or later, the beans will likely not yield as well as when planted earlier.

For the garden farmer seeking only to grow a variety of dry beans for table use, only rather small plots are necessary. You can figure that dry beans will yield at least 30 bushels per acre. A tenth of an acre would then mean about 3 bushels, which will go a long way toward supplying a family with its yearly needs.

Beans do best where soil pH is between 6.5 and 7, in other words, nearly neutral. They like well-drained soil, of course, like most everything does, but I have seen them respond with good yields on some heavy, tight clay soils, too. My grandfather's farm had several low bog-like depressions in some of his fields that had been drained. We called them pot holes. Actually before white settlement, wild cranberries grew in them. Around the edges of these drained bogs was a margin of bluish clay that could all but stop a tractor when we plowed through it. The soybeans always grew best on that blue clay if planting was delayed until the clay was sufficiently dry.

Anyone wishing to grow beans without herbicides should plant in rows wide enough to get a cultivator down between them. They look prettier planted in rows too. I like to plant them thick, about six to eight beans per foot of row, but beans have a way of adapting to seeding rate. Sown thinly, the plants will compensate somewhat and put on more pods per plant than when planted thickly. This is true of many plants, but more pronounced in beans. Planting depth should be about 1 inch early in the season, 2 inches or a little more later in the season when the soil is generally drier.

Commercial farmers in my neck of the woods plant their soybean fields solid, that is, not in rows, because they can control weeds with herbicides. With what are called "Roundup Ready" soybeans (which are genetically modified to be immune to Roundup herbicide), the beans can be sprayed at any time to control weeds without seeming to harm the beans. That's why, for quite a few years now, soybean fields have been so clean of weeds. However, some weeds are becoming immune to Roundup and other herbicides, and the ultimate outcome of this immunity is not yet known. Organic growers who try to plant beans solid run the risk of weedy beans. If you plant late in the season (late June/early July here in northern Ohio) after working the soil well, the beans might stay fairly clean, but don't bet on it. Another argument for planting beans in rows is that even the smallest farmers usually have row planters for corn or vegetables and can use them for beans.

Whether beans yield better planted solid or in rows is a perennial argument among farmers. Apparently solid plantings produce a little more, which is crucial for big commercial growers. For garden farmers the difference may not be enough to offset the high prices of Roundup Ready seed. You can still save your own seed of older varieties for your own use, but agribusiness is trying to patent as many varieties as they can get away with. Many of us contrary farmers believe that the big seed companies push genetically modified seed mostly because they want to force farmers to buy all their seed rather than have the option of saving their own.

Legume seeds are often inoculated with *Rhizobia* bacteria before planting. These bacteria are the agent by which beans can draw nitrogen from the air and deposit it in the soil—the most profound miracle in nature to my way of thinking. Though the bacteria are almost always present in healthy soil naturally, farmers long ago found that adding more bacteria could increase yields slightly. Some scientists think inoculation can be overdone. They say a field that has been regularly planted with inoculated legume seed may not need further inoculation at every planting. Other agronomists disagree, demonstrating that *Rhizobia* populations tend to decline, especially under intensive cultivation and dry weather. The higher the *Rhizobia* population, the better the chances for higher yields, they say. Most legume seeds can be purchased already inoculated. Since inoculation is relatively inexpensive, the extra effort seems worth it, especially for organic growers who are just as interested in the increase in nitrogen fixation that can take place as in an increase in yield.

You can buy special strains of bacteria for various legumes and beans. Most often you find a general inoculant offered for sale for all beans. The inoculant comes in powder or liquid form and can be ordered from most garden-supply catalogs that sell bean seeds. The "powder" or carrier for the bacteria is humus made from reconstituted peat soil. Packed in moisture-sealing plastic, the humus nurtures the bacteria in it and keeps them healthy if the inoculant is stored out of direct sun in temperatures of not over 70°F. Enough moisture remains in the plastic-packed humus to make it stick to the seed when the two are mixed together. Follow application directions on the label. Usually all you have to do is sprinkle the powder over the seeds, then stir and mix in until at least a speck or two adheres to each seed. If the inoculant doesn't want to stick to the seeds, sprinkle a little water over the seeds first, then stir in the powder.

The bacteria will stay alive on exposed seed for about two or three weeks, if it is stored in a cool, dark place. But try to plant the inoculated seed as soon as possible—preferably do the inoculating right before you plant—so that the bacteria get into the ground in vigorous condition.

The soybean, inscrutable as the Orient from which it sprang, holds other secrets, which probably means that other dry beans do too. I found a quotation from a Dr. David Johnson, who probed the secrets of the soybean at the University of Missouri back in the 1970s, to be most interesting. While the soybean does not respond to chemical fertilizers the way other crops do, it seems to be able to find nutrients in the soil other plants can't, says Johnson.

"Research shows that the soybean is able to fix about 100 pounds of nitrogen per acre plus about 50 pounds of nitrogen from its organic matter decomposing in the soil. The other 150 pounds it needs to produce a 50-bushel-per-acre yield, it can get even though no other nitrogen fertilizer has been applied and when apparently there is no residual nitrogen in the soil. It must be getting that nitrogen from the soil, but we don't know how."

Obviously, fertilizing dry beans is a debatable issue. Some chemical farmers claim soybeans respond to application of fertilizer directly, and some claim just the opposite. Many agronomists don't advocate fertilizer on soybeans, contending that the general fertility of the soil and the fertilizer the farmer applied to preceding crops (usually corn, which is heavily fertilized) might help the soybeans, but not nitrogen/phosphorus/potassium (NPK) fertilizer applied at planting time.

On some intensively farmed soils, trace-element deficiency is showing up in soybeans and possibly other dry beans. In this situation, manganese and boron are the two micronutrients the beans are usually lacking. Both deficiencies cause yellowing of the beans or a pale greening. My neighbors add manganese to their fertilizer as a matter of course, applying it to the crop that precedes soybeans in their rotations. In a garden-farm situation, where the soil has not been stressed into trying to produce the very highest yields and where regular green manure, compost, and animal manures are applied, trace-element deficiency should not be a problem. I suppose there are exceptions to that statement, but I haven't run into them in seventy years, so I say stewing over trace elements should be dropped down to about eightieth place on your list of worries.

Weed Control

Obviously, the more weed-free your beans are the better. Mechanical cultivation will keep the row middles clean, but even the most skillful operation will leave a few weeds mingled with the beans in the row. Before herbicides, farmers began cultivation before the beans emerged from the ground. Organic farmers still do. This is accomplished with what is called the rotary hoe pulled at a fairly fast speed behind a farm tractor—as described in chapter 1. If a heavy clay soil has crusted over from a hard rain, rotary hoeing can also break up the crust so the sprouting beans can push through to sunlight.

On smaller garden plots, don't worry about weeds until the beans come up. You can handle them as you do weeds in the vegetables. Begin regular cultivation between the rows with a shovel cultivator or tiller as soon as possible. In the garden with a push-type cultivator or rotary tiller, you can cultivate as soon as the operation does not cover the little plants with dirt. With a tractor cultivator, even with shields to protect the beans from getting covered, you usually have to wait until the beans are about 3 inches high, and then you will have to drive very slowly.

As the beans grow, you should adjust your cultivator so that it rolls dirt into the row around the base of the beans, covering small weeds growing up there, as described in chapter 1. With a tractor-mounted shovel cultivator, you can accomplish this simply by raising the shields slightly and increasing speed. Don't speed up too much, though, because then the shovels will ridge dirt up in the row too high. If the ridge is too high the bean harvester's cutter bar will scrape in the dirt when cutting the beans at harvesttime. Beans, especially soybeans, like to grow pods close to the ground as well as higher up, tempting the harvester operator to cut very low to get all the beans. Then the cutter bar might gouge into the ridged soil, or worse, pick up a rock.

Stop cultivating when the beans are about knee-high, or sooner if few weeds are present. After that time, the cultivator may knock over plants that are beginning to spread out, or injure roots.

Herbicides supposedly have solved the weed problem for farmers, but until the coming of Roundup Ready beans, I can say without fear of too much contradiction that the chemicals usually didn't control weeds any better than cultivation. In those days, I could look out the windows of my home and see acres and acres of soybeans in every direction, all with herbicides applied. All were weedy. The only field of beans in our neighborhood that looked like those beans in herbicide ads—long rows disappearing over the horizon without a single weed in sight—was my Uncle Carl's forty acres up the road. He never used herbicides. But during all of his seventy-odd years, he cultivated assiduously with the tractor-mounted rig and then patrolled the rows all summer, hoeing by hand those weeds that dared to grow back. Farmers from all over stopped to marvel when they passed his beans. They seldom believed us when we said that the only herbicide Carl used was a hoe. He liked to patrol his beans on Sunday after he got home from church. He had a short-handled hoe that he wielded with one hand. Teased about working on Sunday, he would grin and say that using only one hand instead of two was not really work.

Today, as more weeds become immune to herbicides, the soybean fields are starting to look like they used to. Uncle Carl is not around to remind us, but one herbicide that weeds never become immune to is the hoe.

Pests

In the past, insects caused only minor damage to dry beans in this part of the country—another good reason to grow them. But keep your fingers crossed. The Asian beetle, the Japanese beetle, and various cursed leafhoppers are acquiring an interest in beans, especially the soybean, and with the thousands upon thousands of acres of the latter, a population explosion could occur. And that in turn would lead to an epidemic of aerial spraying. This worries beekeepers especially, since dry beans are a good source of honey. Soybean honey is one of the best, and if all soybean farmers kept bees, we could quit using that white, highly subsidized stuff called sugar. (I once had a chemistry professor who often said that "white sugar is poison.") But spraying for harmful bugs could kill the bees. Bean farmers and beekeepers sometimes (not very often) try to avoid the problem by working together. That is, the farmer lets the beekeeper know when he is going to spray. The beekeeper can then shut the bees in the hives on spray day, although this is a real pain in the neck.

Although, having bees myself, I have been very pessimistic about large-scale aerial bombardment of insecticides on soybeans, I must admit that even being next to commercial farmers who sometimes have to do a lot of spraying, I don't think I have lost any bees yet. So I try to remain hopeful. Farmers are becoming more sophisticated too. They are learning not to be too quick on the trigger.

One year we had great hordes of Asian beetles and we all thought the next year would be Armageddon. But the next year there were only a few of the beetles. Natural controls must have kicked in. Perhaps that will happen with the Japanese beetle too.

In the Middle Atlantic states, the Mexican bean beetle sometimes attacks beans of all kinds. But its populations seem to fluctuate from year to year too. In the South, the velvet-bean caterpillar can be harmful, but velvet-beans are not widely grown. Traps and some newer organic sprays help. We have had to spray pole beans for the bean leaf beetle and leafhoppers, but so far not dry beans. Cutworms, white grubs, grasshoppers, and armyworms—the usual meddlesome gang—may occasionally make a raid or two, but seldom present a serious problem.

The worst pests of our edible edamame soybeans that we grow in the garden are deer, rabbits, and groundhogs. Even though they have ten quadrillion acres of soybeans all around us, these animals would rather chomp on the edamames. That should surprise no one. Edamames taste better to us than field soybeans, so why wouldn't that be true for deer? We plant our edamames inside a garden plot surrounded by chicken-wire fence at the bottom to keep out rabbits, and a length of woven-wire fencing above to stop the deer. The latter hangs rather loosely on tall posts, very ugly but cheap. So far so good. Peas and string beans also go inside that fence. Oddly, the deer don't seem to like lima beans beyond a curious taste or two. On small plantings you can drape old curtains or any fine cloth such as the floating row covers now offered in garden catalogs over the plants and the deer are leery. Netting seems to work too. The four-legged bottomless pits don't seem to like nipping down through it.

If some varmint eats off your bean plants early in the growing period, do not despair completely. The plucky bean plant will often grow back again, perhaps with lesser yield, but better than nothing.

Harvesting

Dry beans all form about the same kind of bushy plant resembling the bush string bean. The plants reach as much as waist high in rich ground, though some, like runner beans, may crawl over the soil surface no more than a foot high. Blossoms will form in about two months after the plants come up. Some of the flowers become stunted and fall off, but plenty survive. It is good to have pollinating bees around to get a good yield. Honeybees like bean blossoms.

As the beans mature, the leaves of the plants turn yellow, then brown, and fall off (good for organic matter) and the pods dry and harden. Usually there are three to six beans in a pod depending on species, mostly three. Choose a hot dry period in September or October to harvest when the leaves have all fallen and the beans are hard and dry. Don't wait too long or fall rains might start rotting the beans. (I say that although I know of soybeans that were harvested mechanically in winter because a wet fall prevented earlier harvesting. No statement about farming is completely true.) If I must harvest before beans are completely dry, I spread them out in a very thin layer in the house or barn someplace, or dry them in the stove where you can run the temperature up over about 200° for a few minutes to kill any bean weevil eggs that might be present too. I wouldn't do that for beans that you are saving for seed. A small, solar-heat box, such as you'd use to dry fruit or vegetables in, would make a good bean dryer too.

To harvest, you can pull the plants out of the ground if you have only a few garden rows, or cut off the plants with a sickle bar mower and rake into piles. Then put an armload of the stems and pods in a sack and pretend the sack is that boss who is mean to you. That is, stomp or beat the hell out of the sack. The beans inside will be shattered from their pods much more easily than trying to thresh out grains by flailing. Then you can dump the beans and shattered pods onto a large piece of hardware cloth and shake it. Mostly the beans will fall through and the pods stay on top to be thrown aside. Winnow away any pieces of pod that fall through the screen.

Larger plots are easily harvested with a grain combine, either self-propelled or pulled by a tractor. When I was a boy, my parents would park the Allis-Chalmers five-foot combine and tractor that powered it next to their garden patch of navy beans and while it ran in a stationary position, they would fork the bean plants into it. The combine and tractor were too big to get into and out of the garden.

If you have an in-between-size planting—more than a garden plot, but not really enough for a field harvester—I have a notion that a lawnmower might thresh out beans fairly well. (That is, I wish someone else would try this, because I don't have a bagger attachment on my mower.) I know the mower will thresh beans because it will thresh out wheat, which is harder to do, but whether the beans will blow on into the bagger or will just drop on the ground, I don't know. Some of the beans would split in the process, but that's true of a field harvester too. And split beans are OK for many dishes. Split pea soup, for instance.

Sometimes you will want to eat your dry beans more as a green garden vegetable—just as you can let string beans mature for baked beans, which we used to do occasionally. (They aren't as tasty, though, as navy beans or other beans normally used for baking.) For example, we eat edamame soybeans green, plucking the pods and steaming them. With a little practice you can learn to squeeze on the pod and pop the steamed beans right into your mouth. Very cool to do in a ritzy restaurant where steamed beans are served.

There are a zillion cookbooks now to tell you how to make and use soy milk, bean sprouts, and many prepared foods from dried beans. There is vigorous debate over the benefits of soy milk, but that is something for nutritionists to argue, not me. Vegetarian cookbooks have the most information. I would just underline the fact that grain flours and bean flours can be mixed into breads and other foods to great dietary advantage. A few recipes at the end of this chapter give you an idea of the possibilities.

Being a lover of baked beans, I think everyone should keep a bag of dry beans in the basement where it is cool and dry. They store very nicely that way and seldom are bothered by bugs or mice like wheat can be. My mother always had a sack of navy beans hanging from the stair steps leading into the cellar. We could hardly starve because we could eat beans all winter long. She baked them with a layer of bacon from the smokehouse on top, and oh my, a poor man could feel like a king.

Thinking Outside the Bean Bag

Writing about the partnership of grains and beans tempts me to a far-out proposition. I pose it only for your contemplation and inspiration, because, right now at least, basic foods like grain and beans are rising dramatically in price. Suppose, just for the fun of it, that a garden farmer wants to get into the company of the big boys—big tractor farmers—but can only afford a hundred and twenty acres. With farm prices skyrocketing in 2008, it is tempting to believe that even at such a comparatively small size, you might make a little money in your spare time. Let us say the garden farmer already has a smaller farm tractor and the tilling equipment to plant a hundred and twenty acres, and an older, small grain combine to harvest it; or can hire one of the big boys to harvest his crops for him. If I were in this situation, that is, younger and more energetic than I am now, here's what I would do.

This is not originally my idea. When I was employed by the Soil Conservation Service several eons ago, a co-worker and I used to while away more time than we should have dreaming up "perfect" farming systems—perfect in the sense of practicing really sustainable soil conservation and making a nice profit at the same time—which often seems to be something of an oxymoron. One of our favorite such systems was inspired by my Uncle Carl, whom you met earlier in this chapter—the guy with the one-handed hoe. Carl in his later years was seized with a strange idea. He quit growing corn. To quit corn in the Corn Belt is tantamount to losing one's mind in our community. Carl did not mind being scorned because he had his own peculiar outlook on everything. Once when we were discussing the sudden upswing in the number of swimming pools in town, he opined: "Yep, and most of those people don't have their bathtubs paid for yet." So anyhow, he quit raising corn because, he said, the cost of putting out the crop was about the same as the income from it, which was mostly true, especially for him. He would not spend the money to sock on heavy applications of fertilizer to make the corn yield more bushels. His reasoning was that in years when the weather was less than optimum, he wouldn't get the yields that the heavy fertilization promised, which is true. All that extra expense down the drain. He reminded me of another farmer who decided one year that he could make money on a field of oats, also contrary to local wisdom. He risked fifteen acres and figured later that he lost only $20 an acre. "Lucky I didn't plant a hundred acres, isn't it?" he remarked.

Anyway, Carl went to a simple rotation of soybeans and wheat on his farm, the beans not needing any fertilizer and in fact providing, by nitrogen fixation, all of that precious nutrient that he figured the wheat needed.

So, while we were admiring Carl's independent thinking, my coworker proposed a tweak to my uncle's system that made the rotation, at least in our eyes, "perfect."

Bear with me now. I hate the word "scenario," but here's what we came up with—on paper. Between the soybeans and the wheat, we inserted a crop of red clover. That meant that the rotation would be wheat, sown in the fall, clover seed broadcast-sown on the wheat the following spring, the wheat harvested in July, the clover coming on strong by fall to be harvested for the seed. The next year a cutting of hay would be taken, and clover seed again harvested in the fall from the regrowth. In the third year the field would be plowed and planted to soybeans, which would be harvested in the fall and the land sown immediately back to wheat. And so on over and over again until death do us part. All of these practices are tried and true—nothing revolutionary except the absence of corn. Even harvesting clover seed was often done in those days, and I think it is a shame that most farmers don't do it anymore. Anyway, here's what we were looking at: On, say, a 120-acre farm, 40 acres would be harvested for wheat, straw, and clover seed; 40 acres would be cut for hay and the regrowth harvested for clover seed again; and 40 acres would be harvested for soybeans and planted back to wheat.

The 40 acres of wheat, yielding 60 bushels per acre at early 2008's unreal price of about $7 a bushel (the price has fallen at least temporarily in late 2008) equals $16,800. The wheat straw, baled after the grain is harvested, should make about 2 tons per acre valued at say $60 a ton or $120 per acre, or $4,800 for 40 acres. The first harvest of clover seed should amount to about ½ bushel per acre, or 20 bushels at a value of $60 a bushel, or something like $1,200. (It could be worth $100 a bushel sometimes.) The year-old clover in the next field should make 2 tons of hay per acre, and the price, depending on quality, could easily be $150 a ton right now, or $300 an acre, or $12,000 for 40 acres. The clover seed harvested in the fall of the second year should make a bushel per acre or about $2,400 for 40 acres. In the third field the soybeans ought to yield at least 40 bushels per acre or 1,600 bushels or, at early 2008 prices, $15 a bushel, (down some by late 2008 prices too) or another $24,000. That all adds up to $61,200, if I can still do the arithmetic without a calculator. Now you see why farmers were acting so giddy in mid-2008. Who knows? Maybe such prices will continue. (I don't think so, but what do I know?)

The catch here is that grain and bean prices are twice as high as normal and so are expenses. No one knows how long this situation will last, but if it does, land prices would no doubt rise to $10,000 an acre or more (an Illinois grain farm sold for farmland in October 2008 for $9,000 an acre), and if you have to borrow that kind of money, you are back to square one again. *But*, if you have the land, either paid for or *being* paid for with funds outside of farming, *and* that land is reasonably fertile, *and* you follow this low-cost rotation, *and* do not use any purchased fertilizer, *and* you farm with horses or old tractors, *and* have family help to bring in that hay, you *could* have a nice little profit. Even if things return to more normal conditions, your gross could be around $30,000 on these 120 acres, and you could figure a net of around $8,000—again if you farmed without purchased fertilizer and with horses or your old, paid-for tractor and machinery. Not bad part-time income.

Bean Recipes

What follows are a few ideas and recipes for preparing soybeans and other beans, just to give the newcomer to bean foods an idea of what is possible. There are now many, many cookbooks to show you how to become a gourmet of the poor man's meat.

Cooking Soybeans

- Wash soybeans, then remove any foreign particles. Cover soybeans with cold water. Refrigerate or freeze overnight. Freezing will lessen the amount of cooking time needed.

- Next day put the soybeans and their soaking water on to cook, using a large enough pot and leaving the lid slightly to one side so that the soybeans will not boil over. Bring soybeans to a boil, then turn heat down and simmer until they are tender, 2 to 3 hours. (They will never be as soft as navy beans, for example, but they will get tender, when done.)
- Soaking or cooked soybeans ferment very quickly, so should never be left very long at room temperature. They will keep quite long in the freezer, but not more than a few days in the refrigerator.
- One cup of dry soybeans will swell during soaking to 2½ to 3 cups. One pound of dry soybeans is about 2¼ cups.

Basic Soy Milk

⅔ cup soybeans
4 cups water
⅓ cup oil
¼ teaspoon salt
¼ cup honey

- Soak soybeans in cold water overnight. If the weather is hot, refrigerate them.
- Drain soaked soybeans and discard water.
- Put soybeans and 2 cups water into container of an electric blender. Blend at medium-low speed for about 3 minutes.
- Put soybean mixture into the top of a double boiler. Stir in 2 more cups water and cook over rapidly boiling water for 30 minutes.
- Strain soybean mixture through a cheesecloth-lined strainer. Rinse out top of double boiler and return strained soy milk to it. Cook 30 minutes over rapidly boiling water.
- Strain soy milk through cheesecloth-lined strainer.
- Pour ⅔ cup hot soy milk into an electric blender, add oil, and blend on medium-low speed for 5 minutes. (The full time is necessary in order to obtain an emulsion.)
- Add remaining soy milk, salt, and honey and blend about 2

minutes longer. Cool. Store in a covered jar in the refrigerator. Shake vigorously before using. Keeps well for a week.

**Note: Use as you would milk, over cereal, in cooking, and beverages.*

Yield: about 1 quart

Blender Soyburgers

1 medium-small onion
4 tablespoons vegetable oil
2 cups soybeans, cooked
1 cup diced potato, cooked
¼ cup fresh bean sprouts
1–1½ teaspoons sea salt
½ teaspoon dried dill weed
1 teaspoon dried parsley
1 teaspoon dried basil
Few dashes of powdered sage
1 teaspoon oregano
Vegetable oil

- Peel, chop, and grind the onion in a blender at medium speed. Leave in the container. Add 2 tablespoons vegetable oil and 1 cup soybeans, a tablespoon at a time through the blender cap. Scrape out into a bowl. Grind the rest of the beans with another 2 tablespoons of vegetable oil, and scrape out.
- Drop half the potato into the blender and whirl at low speed until it forms an elastic mass; scrape this mass over the beans.
- Repeat. (Cooked potato doesn't mash in the blender, it forms this starchy elastic stuff that will help hold the burgers together.)
- Add the sprouts, salt, and herbs, and stir well. If you have the time, let stand for half an hour, to allow the herbs to soak in. Taste for salt.
- Spoon onto a lightly oiled baking sheet and broil (starting with a cold oven) at highest heat for 15 minutes on the first side and 5 minutes on the second.

Yield: 10 to 12 large burgers

Barbecued Soybeans

6 cups cooked soybeans (approximately 2¼ cups uncooked)
The water soybeans were cooked in (or the equivalent amount of vegetable stock)
½ cup onion, chopped
1 clove garlic, minced
2 tablespoons oil
½ cup tomato juice
½ cup catsup
1 tablespoon molasses
½ cup green pepper, chopped
2 tablespoons honey
1½ teaspoons dry mustard
¼ cup parsley, freshly chopped
⅛ teaspoon cayenne pepper

- Preheat oven to 350°F.
- Put cooked soybeans with their cooking liquid or stock into a lightly oiled casserole.
- Sauté chopped onion and minced garlic in 2 tablespoons oil until golden but not brown.
- Remove from heat and add to cooked soybeans. Add the tomato juice, catsup, molasses, chopped green pepper, honey, dry mustard, chopped parsley, and cayenne pepper. Mix together thoroughly.
- Place in oven and bake 1½ hours or until soybeans are tender.

Yield: 6 to 8 servings

Swedish Soybean Soup

2 cups soybeans
Water to cover
1 medium-sized meaty beef bone
3 quarts cold water
Salt to taste
½ teaspoon paprika

1 cup chopped celery, with leaves
1 cup onions, chopped
3 medium-sized turnips, diced
½ cup parsley, chopped
⅛ teaspoon cayenne pepper
1 cup tomato puree or canned tomatoes
Chopped parsley for garnish

- Wash soybeans and discard any beans with imperfections. Cover soybeans with water and place in refrigerator, covered, overnight.
- The following day, place soaked soybeans in a large, heavy soup kettle. (Be sure to use a large enough pot and leave partially uncovered, so as to avoid soybeans cooking over. This can happen very easily.) Add beef bone and 3 quarts of cold water. Place over medium heat and bring to a boil, uncovered, removing any foam from surface as it accumulates. Reduce heat; add salt and paprika. Cover partially and allow to simmer for 3 hours, stirring occasionally.
- Add the chopped celery, onions, turnips, parsley, cayenne, and tomato puree or canned tomatoes. Cover partially and allow to simmer for another hour or until soybeans are tender. Continue to stir occasionally while cooking. Taste and correct seasoning. Garnish with chopped parsley.

Yield: approximately 3 quarts

Orange Muffins

⅔ cup oat flour
½ cup brown rice flour
⅓ cup soy flour
½ teaspoon salt
3 tablespoons wheat germ
2 teaspoons dry yeast
2 egg yolks
3 tablespoons honey
2 tablespoons oil
¼ cup nonfat dry milk

1 cup water
¼ teaspoon mace
1 tablespoon orange rind, grated
¼ cup raisins
2 egg whites

- Prepare a 12-cup muffin pan by brushing with oil (2½-inch-size cups).
- Sift together flours and salt into a medium-sized bowl. Add wheat germ and dry yeast; mix together.
- In a mixing bowl, beat egg yolks until thick. Add honey and oil. Combine nonfat dry milk and water with a wire whisk, and add to egg mixture. Stir in flour mixture. Add mace, orange rind, and raisins.
- Beat egg whites until soft peaks form. Gently fold beaten egg whites into batter until well combined.
- Spoon batter into prepared muffin-pan cups, filling two-thirds full. Place in a warm area or over a shallow pan of hot water, cover, and allow to rise for 30 minutes. Meanwhile, preheat oven to 375°F.
- Place raised muffins on middle of rack in oven and bake 30 minutes or until nicely browned.
- Remove from oven and loosen edge of each muffin with a spatula: remove and serve immediately.

Yield: 12 medium-sized muffins

Ground Beef Stuffed Peppers (with Soy Grits)

6 medium-sized green peppers
½ cup onion, chopped
1 clove garlic, minced
¼ cup olive oil
1 pound ground beef
½ cup soy grits (cracked dried soybeans)
1 cup brown rice, cooked
1 teaspoon tamari soy sauce
½ teaspoon salt
1½ cups tomato juice

¼ cup parsley, chopped
3 tablespoons wheat germ
3 tablespoons grated Parmesan cheese

- Preheat oven to 350°F.
- Wash and drain green peppers. Remove stem ends from peppers and carefully remove seeds and ribs. Set aside.
- Sauté onion and garlic in olive oil until tender. Stir in the ground beef and continue to sauté only until it is no longer red. Add the soy grits, cooked rice, tamari soy sauce, and salt. Mix together to blend ingredients. Stir in ¼ cup tomato juice and the chopped parsley. Cook for 5 minutes.
- Remove from heat and allow mixture to cool slightly. Stuff the peppers with the meat mixture. Place in a baking dish and pour 1 cup of tomato juice around the stuffed peppers. Cover and place in oven and bake 45 to 50 minutes or until peppers are tender. Baste occasionally with pan liquid. Remove cover after the first 30 minutes of baking. Top peppers with wheat germ combined with grated Parmesan cheese, and brown slightly.

Yield: 6 servings

CHAPTER SEVEN

Rye and Barley

Rye

Rye is not a particularly impressive grain either in terms of yield or nutritional value, except perhaps to lovers of the whiskey distilled from it. But rye's popularity continues because of its amazing tolerance for cold weather, which makes it an ideal winter cover crop and a desirable pasture plant for livestock in fall and spring, when other pastures have quit or just started to grow. Rye can actually begin to germinate when the air temperature is in the 30s on soil warmed by the sun to something higher than the air temperature. This is an interesting phenomenon to study with a thermometer. I have occasionally noted in winter that not only rye but bluegrass and red clover will grow ever so slightly when the air temperature, say in a January thaw, is barely up to 40°F as long as the sun is shining. The sun's rays can warm the soil-surface temperature on such a day up to 45° to 50°, which initiates a little growth. (I knew a farmer in Minnesota in my wild oat days there, who took advantage of this phenomenon and planted radishes on a steep southern slope very early in the spring when snow might still be present in woodland shade. He had radishes at the farmers' markets in Minneapolis just as early as radishes shipped in from farther south.)

Once established, rye will continue to grow in the fall until the temperature drops below 40°F, and resume growth when the temperature rises above 40°F in the spring. It is the hardiest of all grains and won't winterkill at 40 below zero. On the other hand, rye doesn't like hot weather and won't germinate very well when the temperature is above 85°F.

Rye's second advantage is that it will produce a crop on land too poor for wheat. In the cold climate of northern Europe, where soil is poor, rye is an important and dependable source of bread flour.

Rye

In the United States, rye is more important as a cover crop, or for green manure and pasture than as grain. The crop can, however, be grazed awhile in the spring and then allowed to grow again for harvested grain. Some dairymen like to grow a little rye strictly for late fall and early spring pasture, then plow under the residue. Organic market gardeners grow it for winter cover and then incorporate it into the soil as a green manure with a heavy rotary tiller. If allowed to mature, rye grows much taller than other small grains, taller than a man. Thus the old lovers' song about "if a body meets a body, comin' through the rye." Seems like a lot of hi-liggety was going on even in the olden days when people were supposed to be oh-so-much more proper than today. Just one more piece of evidence that farming is not, and never was, a boring occupation.

Grow rye just as you would winter wheat. It can be sown any time between late summer and late fall, preferably during the earlier half of that period so that it can be pastured in the late fall. Later plantings, for spring plow-down or grain, should not be postponed so long that they can't establish themselves before cold weather arrives in force. In Ohio, a planting for grazing should be completed ideally by September 15, and for grain before the end of October. Adjust those dates to fit your own climate.

For grain, plant about 1½ bushels of seed per acre; for grazing, 2 to 2½ bushels per acre. Plant either with a drill or by broadcasting as described in chapter 3.

Rye will ripen just ahead of winter wheat. It yields poorly by comparison, perhaps 30 bushels (or less) to the acre, whereas wheat will produce 50 bushels per acre or more. That means there is not much economic incentive to grow rye for grain if you can grow wheat unless you have a special market for it as seed, or can sell it to distilleries or to horse farms. In the latter case, horse owners sometimes prefer rye straw to wheat for bedding if they can get it. Cut when the grain is in the soft dough stage, the straw is brighter and, of course, longer than wheat straw. Rye straw looks showier and is not apt to cling to the horse like shorter wheat straw might.

Rye grain is not as palatable as other grains for livestock, and it's harder for animals to chew. Its chief value for garden farmers is for making tasty homemade rye bread. You can grow all you need for that in a garden patch. Harvest just as I have described for wheat, and grind in a blender or mill. Mix rye flour about half and half with wheat flour for rye bread. Most Americans prefer it that way anyhow. Rye seed is available from some mail-order seed catalogs, like Shumway and Seeds of Change.

A new development in rye works to the benefit of the organic gardener using it as green manure. Tetraploid rye—rye with double the number of chromosomes as ordinary varieties—was perfected first for ryegrass, which is not the same plant as rye grain. (Tetraploid ryegrass is now common, a favorite pasture grass of many grass farmers, but with no grain value.) The tetraploid characteristic gives the plant more vigor and, hence, faster growth and better yields. Michigan State released the first tetraploid rye in 1973, calling it 'Wheeler' rye. In addition to being an excellent green-manure crop, it is considered a source of high-quality silage, which most ryes are not. Tetraploid ryes should not be grown with other varieties. Cross-pollination may cause sterile seed. That's what happens when you start monkeying around with chromosomes.

Farmers concerned with soil erosion (unfortunately not all farmers are) will sometimes broadcast rye in standing corn in late summer, often by airplane. Enough sun can get through the corn as it matures and dries to allow the rye to grow and cover the bare ground between the rows with a carpet of green. By the time the corn is ready for harvest, the rye not only is helping to control erosion but also is providing a firmer soil surface for machinery when the ground is muddy. The following spring, the rye is plowed under for green manure.

You can follow the same procedure in your garden, as many organic gardeners do. Rye sown between rows of late fall vegetables will keep you from having to wade through mud from late fall and early winter rains. Before it grows tall the following spring, mow it with a rotary mower and till in the clippings.

Insects and diseases

Insects that attack other cereal grains will occasionally show an interest in rye, but rarely to any serious extent. Rye is a good organic crop for that reason. It is, however, susceptible to ergot disease, a fungus that produces black growths called sclerotia that replace the kernels in the rye heads. Ergot is poisonous to humans and livestock. There is a curious theory in history that claims the

French Revolution turned so violent because the people had been poisoned and driven half-insane by an epidemic of ergot in the wheat that year. Sounds like a theory espoused by the nobility.

All rye used for seed should be clean and ergot-free. If you use seed that is over a year old, you may automatically control ergot because the sclerotia lose viability after a year's time and hopefully won't carry the disease over to a new crop. Ergot can be removed from grain by soaking infested rye in a 20 percent solution of common salt in water. Stir until the ergot bodies float to the surface and skim them off. You'll have to wash the salt off the seed before planting it. (I label this whole business NWTE, that is, not worth the effort. Throw ergot-infected rye away.)

Rye is afflicted with certain stem rots and smuts, but crop rotation and newer, resistant varieties solve that problem. The smut spores will not survive in soil much beyond a year, so rotations that space rye plantings at longer intervals than one year should avoid infestations.

Anthracnose is a problem, especially in the humid South. The disease is associated with poorer soils and seldom is severe on organic soils where well-balanced fertility is maintained.

Rye Recipes

Sourdough Rye Bread (Wheatless)

1 cup sourdough starter (see instructions below)
1½ cups water
2 cups rye flour
1 tablespoon molasses
½ cup lukewarm water
1 tablespoon dry yeast
2 tablespoons oil
2 teaspoons salt
4 cups rice flour
2 cups rye flour
⅓ cup (approximately) oat flour (oatmeal coarsely ground in electric blender)

To Make and Maintain a Sourdough Starter:

- Dissolve one package of dry yeast in ¼ cup warm water. In a large ceramic or glass bowl combine 1 cup milk, 1 cup all-purpose white flour, 1 cup sugar, and the dissolved yeast. Using a non-metallic spoon, beat the mixture until smooth. Then stir in another cup of milk and flour. Cover the bowl with a clean kitchen towel and set aside at a warm room temperature for 24 hours. After this time, transfer the starter to a covered glass or plastic container and store in the refrigerator. Stir once a day for the next five days.
- Whenever you want to use the starter, remove the amount you need for the recipe and allow it to come to room temperature. Replenish the starter by adding equal parts flour and milk, and half a part of sugar.

To Make the Bread:

- Preheat oven to 375°F.
- To make the "sponge," combine the first three ingredients in a large bowl, cover, and set in a warm, draft-free place overnight. (A cold oven is a good place.)
- Next morning, dissolve molasses in lukewarm water and sprinkle yeast over the surface. Set aside for 5 minutes to activate. Stir down the "sponge," add oil, salt, activated yeast mixture, and all of the rice flour.
- Mix in as much of the remaining rye flour by hand as possible, then turn out dough onto a board or counter that has been well floured with the rye flour. Knead dough briefly, to incorporate all the flour and finish with the oat flour to reduce stickiness of dough. Place in an oiled bowl, turning the dough to oil its surface.
- Cover and put in a warm place to rise for about 1 hour, or until double in bulk. Form dough into one large loaf and one small loaf, place in well-buttered bread pans, and leave to rise to the top of the pans (approximately 1½ hours). Bake in oven for 35 minutes or until done. Remove loaves from pans and cool on rack.

Yield: 1 large and 1 small loaf

Herbed Batter Bread (An All-Rye, No-Knead, Casserole Bread)

¼ cup nonfat dry milk
1 cup lukewarm water
1 tablespoon honey
1½ tablespoons salt
1 cup lukewarm water
4 teaspoons dry yeast
3 tablespoons parsley, chopped
2 tablespoons chopped fresh basil, or 1 teaspoon crushed dried basil
1 tablespoon chives, freshly snipped
1 teaspoon crushed dried oregano
½ teaspoon fresh chopped thyme or ¼ teaspoon dried thyme leaves
½ teaspoon freshly snipped marjoram or ¼ teaspoon dried marjoram leaves
4½ cups rye flour
2 tablespoons oil

- Combine nonfat dry milk and warm water in a small bowl, using a wire whisk. Add honey and salt.
- In a larger mixing bowl, place the lukewarm water and sprinkle the yeast over the top, stirring until dissolved.
- Add the milk mixture to the dissolved yeast. Stir in the fresh and dried herbs. Add 2 cups of the rye flour. Beat for 2 minutes on medium speed in an electric mixer, scraping the sides of the bowl frequently; or beat vigorously with a wooden spoon, about 200 strokes, until the batter looks satiny. Using a wooden spoon, blend in the additional 2½ cups rye flour.
- Scrape batter from the sides of the bowl. Cover with a clean towel and set in a warm place (85°F), away from drafts, to rise until light and double in size, about 45 to 50 minutes. (Do not allow batter to over-rise.)
- Stir the batter down. Turn into a well-oiled 1½-quart casserole or soufflé dish. (The batter will be sticky. Smooth out the top of the loaf by patting into shape with a floured hand.) Again, allow to rise in a warm place, covered, for 20 minutes. Preheat oven to 375°F.

- Place in oven and bake for 45 to 50 minutes or until bread is golden brown.
- Remove from oven and brush top of bread lightly with oil. Cool for 10 minutes. Turn bread out onto wire rack to cool.

Yield: 1 round loaf

Rye and Lentil Pilaf

1 small onion, minced
½ cup diced carrot
½ cup diced celery
2 tablespoons oil
1½ cups cooked whole or cracked rye (approximately ½ cup uncooked)
1½ cups cooked lentils (approximately ⅓ cup uncooked)
1 teaspoon caraway seeds (optional)
1 teaspoon thyme
¼ teaspoon sage
¼ to ½ cup chicken stock or tomato juice
Salt and pepper to taste

Sauté onion, carrot, and celery in oil until tender. Combine with cooked rye and lentils, add herbs and chicken stock or tomato juice. Season to taste, cover, and steam for 10 minutes, or until hot.
Yield: 4 servings.

Barley

If barley isn't the oldest cultivated grain as some historians believe, it certainly is the most widely adapted. Barley can grow inside the Arctic Circle and barley can grow in the tropics. There are varieties suited for Montana and Arizona, New York and Georgia, Tennessee and Minnesota. And all points in between. Barley likes a cool ripening season and moderate moisture best, but will adapt to hotter conditions. Since it uses most of the water it needs during the winter and spring months and can then endure dry weather

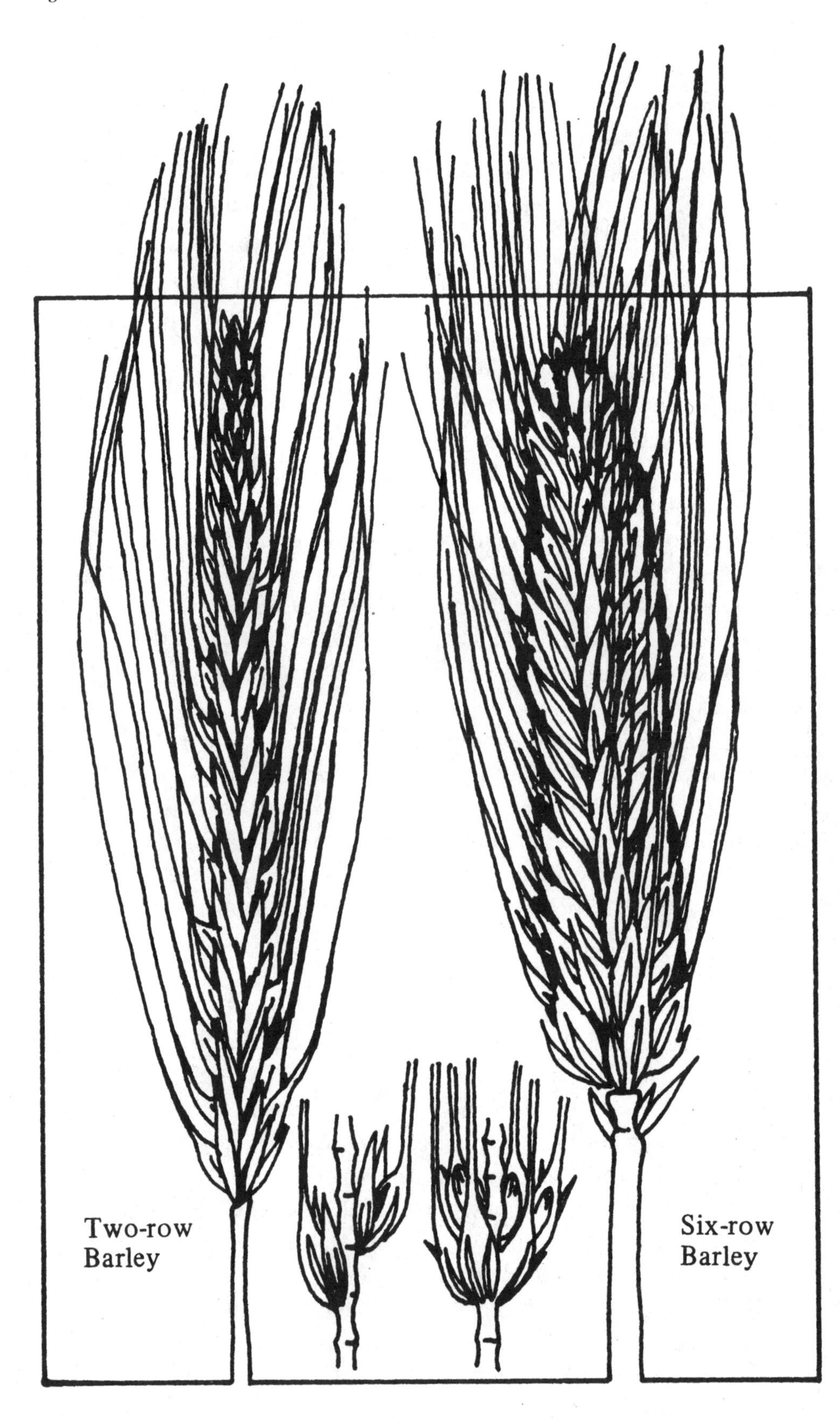
Two-row
Barley
Six-row
Barley

until harvest, barley is a welcome crop on irrigated land to rotate with cotton, alfalfa, sorghum, and other summer-growing crops in the South. Moreover, barley is more tolerant of salt than other small grains and adapts somewhat to saline western soils.

Barley makes good feed for all livestock. It contains almost as much energy as corn and three percent more protein on the average. It can replace corn pound for pound in cattle rations. In the Northwest beyond the Corn Belt, where the largest barley acreages are found, beef and pork are traditionally produced using barley feed instead of corn. Northwesterners believe that barley imparts a distinctive flavor to the meat, which they prefer to corn-fed beef or pork.

However you like your steaks, barley makes good soups and cereals in the hulled (pearl barley) form. Alone, barley does not have enough gluten to make bread and must be mixed with other flours. Sprouted barley, dried and crushed, is the principal malt used in brewing beer, distilling Scotch whiskey, and making malt syrup for other food purposes. Some 30 percent of the barley grown is used for malting.

To achieve so much versatility with barley, good old *Homo sapiens* has developed varieties from two botanically distinct types of barley: six-row and two-row. The six-row varieties are more common, and they are divided into three distinct families. Group one contains the malting barleys of the upper Midwest: tall, awned (bearded), spring-planted, and tracing back to Manchurian origins, the name by which the group is sometimes identified. Group two is called the coast group, the barley usually grown in California and Arizona as winter or fall barley. It is North African in origin. Group three is the Tennessee winter group, the true winter barleys grown east of the Mississippi, principally for livestock feed.

Two-row barleys are grown in the Pacific Northwest, the Intermountain region, and the northern half of the Great Plains. They are spring-planted both for cattle feed and malting.

Because of the different purposes for which barley is grown, there are many varieties, and new ones are offered continuously. In most cases, breeders are seeking a stiffer-strawed plant. You can find the latest sources on the Internet or talk to the manager of your nearest grain elevator and farm-supply store. Hybridizing barley for higher yields (actually to give seed companies more control over the market, in my ornery opinion) is an on-again, off-again proposition. It is doubtful that hybrids will ever yield significantly higher than non-hybrid varieties, which can produce 80 to 100 bushels per acre.

Barley varieties may be beardless, but usually sport beards like young men are doing today. The "beards" are slender bristles, usually about 3 inches long, that grow from each seed in a head and are more correctly called *awns*. Farmers who grow barley for forage generally prefer beardless varieties, since horses may not eat the awns, especially if they are rough or barbed, as is often the case. But, for that very reason, bearded varieties have been found to be resistant to deer foraging. The deer don't like the beards either.

I have my own small adventure to tell about bearded grain (some wheat varieties are bearded, too) and to pass on as a word to the wise. In my wandering Minnesota days, I worked several summers on farms where grain was still threshed the old-time way, with a threshing machine. Despite the scratchy straw and rough, bearded grain heads, I worked the first day shocking the bundles without a shirt on. Big he-man stuff. Each shock, when finished, had to be "capped" to keep out rain. The cap was just another bundle fitted over the shock to make an inverted V "roof." To bend the bundle, or break it as we said, the shocker had to clasp the butt end against his stomach with the left arm and bend the straws over that arm with the right hand, almost as if he were folding a blanket. The bundle could then be set over the shock.

To make a long story short, that night I woke up with a nagging pain in my stomach. By the third awakening, I decided that I had been stung severely on the navel, though I could find no break in the skin. The fourth time I awoke from a nightmare in which I had been stabbed in the stomach. Close examination revealed that a barbed awn was literally working its way inside me via the navel. The barbs of the awn faced outwards: the awn could slide forward into me with the movement of my torso, but the barbs prevented it from sliding backwards. To this day I wonder how far that awn would have penetrated. Moral: When shocking grain, keep your shirt on.

Grow barley almost exactly the same way you'd grow wheat. Some is spring-planted and some fall-planted. Fall-planted winter barley may winter-kill where winter temperatures average less than 20°F. As a general rule, plant winter barley where winter wheat is planted; spring barley where spring wheat is grown. Fertility requirements are similar. Barley ripens sooner than wheat: spring barley in sixty to seventy days, winter barley about sixty days after growth begins in the spring. Because barley ripens quicker than wheat, it fits into a double-cropping system better. A second crop planted after barley has longer to mature than when planted after wheat. The garden farmer seeking to make a small acreage as productive as possible could plant soybeans after barley even in the North, or sorghum after barley in the South, with a good chance of success.

The first experience I had seeding barley brought me into conflict with those blasted beards again. I was just a boy at the time, and my father was planting barley from which the awns had not been removed. The awns caused a problem in the old drill we were using. They'd bridge over and plug the tubes that fed the grain from the planter boxes down into the ground. It fell to my unlucky lot to keep all eighteen tubes open so that the drill planted the seed evenly with no missed spots. That was no easy job, and that's why I remember it so clearly. I doubt you'll come across seed as rough as that today, but if you do, don't try to run it through a planting drill.

In the field or in the garden you can put barley in your rotation as a replacement for wheat. Barley rather than corn is a good crop to follow potatoes since scab disease in potatoes can carry over on corn but not on barley, I understand. Where "take-all" disease (already mentioned in chapter 3) became serious in wheat fields in California, farmers began planting barley instead, since the disease doesn't attack barley so seriously. In dry climates, a rotation of sorghum, barley, fallow, and wheat has been found profitable. Sometimes the barley is grown right on the summer fallow. In the South, barley is sometimes rotated with cotton (this was especially true in the past). The cotton yield is improved, and the barley lessens winter and early spring wind erosion.

Experiments with commercially grown, irrigated barley in Arizona have applications for garden farmers. Researchers there have found that barley grown in rows 14 inches apart produced as good a crop as solid-planted barley, even though less seed was used. The plants in rows had more room to tiller, and the greater number of tillers not only improved yield, but anchored the plants better and kept them from lodging. What's more, research reported much greater efficiency in fertilizer use, and weeds could be controlled by cultivation in the early stages of the crop.

For garden farmers, planting small grains in rows, however untraditional in America, makes sense. They, too, can use less seed, can use precious compost and organic fertilizers more efficiently, and can cultivate for weed control instead of using herbicides. Also, for hand harvesting, grains in rows will make bunching the cut stalks into bundles easier.

Feeding Barley to Livestock

Barley can be fed whole to rabbits, chickens, hogs, or livestock. Chickens don't seem to like it as well as wheat on account of the hulls, and all animals will consume more if the barley is ground. Better than grinding, I think, is to sprout the barley, at least for feeding to a small number of chickens or other animals. After all, most organic growers are already well aware of the value of sprouts in their own diets. Sprouted oats and other grains were standbys for chicken feed in the good old days. Why not utilize barley that way now, since its sprouting ability is so well documented? The sole hitch is that you must store the barley four to six weeks before trying to sprout it. The process is called "after-ripening," and it is necessary for prompt germination. With the new interest in vegetable juices, barley-grass juice from sprouts is now considered, like wheatgrass juice, to be extremely healthful.

An easy way to sprout small amounts of barley for a few farm animals is to soak the grain heads when they are still attached to the stalks in tied bundles. Do one bundle at a time. It should sprout in about five days at a temperature of 60°F. Eventually, you'll know how much barley to keep in the soaking process so as to keep a steady ration coming every day. The barley will sprout right in the head, and you can toss the whole bundle or part of it to the chickens. They get excellent feed, the stalks make excellent bedding, and you don't have to thresh the barley.

Malting Barley for Beer

I have never made beer myself, but from remembering my father doing it years ago, and then listening to the tales my father-in-law told (he was a successful moonshiner) here's roughly how to do it. (More detailed instructions can be found in many books, including my own *Good Spirits*, published in 2004 by Chelsea Green Publishing.)

First you sprout your barley. (You can avoid the whole sprouting routine by buying malt extract). Doing your own sprouting is not difficult if you've had any experience with other grain sprouting. You need to keep the grains moist but not let them mold, at a temperature of around 65°F. Any way you accomplish that will suffice. Distillers of Scotch whiskey used to wet down a layer of grain about 6 inches deep, and keep turning it over manually with a shovel. You allow the barley to sprout only until the plant sprout—not the root sprout—is two-thirds of the length of the grain itself. The sprout will not have emerged yet, but is plainly visible in the swelling grain. As soon as it reaches the prescribed length, which should be in about ten days, dry the grains at a temperature of 130°F, and never more than 140°F. When dry, the grains should be brittle and crack sharply between your teeth.

To make beer, you need at this point some kind of barrel or container (wood or copper in the old days, though I think food-grade plastic would be okay if it is the kind that can take boiling water) that has an outlet near the bottom. A wooden cider barrel with a spigot near the bottom would be fine.

First you crack your malted barley. A coffee grinder, blender, rolling pin, or roller mill will do the job. Crack the grain, don't grind it to a powder. Next, mix water with the cracked malt to make a thoroughly wet, heavy mash. The water needs to be boiled and then cooled to 150°F before using it to make the wet mash. Let the mash stand overnight.

Next morning, pour boiling water on the mash. The water soaks down through the mash and is drawn off through the bottom spigot or whatever arrangement your ingenuity has devised. Your formula to follow is 10 gallons of boiling water to 1 bushel of mash, to which you will later add 1 pound of hops if you are making beer. So ½ bushel takes 5 gallons of water and ½ pound of hops, and so forth.

After you have poured the proper amount of boiling water through the mash and drawn off the liquid, the latter is strained to remove any pieces of grain in it and then boiled with the hops, the hops contained inside a cloth bag in the liquid. The mash left over is still very nutritious feed for livestock, something that is more important to me than the beer. Well, maybe.

Boil your beer and hops for about an hour. Then pour the liquid into another container and get it cooled down as fast as you can. If it cools too slowly, it may spoil. A good way to cool it quickly is to run cold water through a coiled copper pipe immersed in the beer. (Cooling in a bulk milk tank would be perfect, but don't let your milk inspector know.)

While the beer cools, take a gallon of it and cool it down quickly to body temperature; then add yeast, and let it work. When the rest of the brew has cooled to about 60°F, dump in the gallon of yeasted liquid, cover the whole with a cloth so insects can't get to it, and let the brew ferment for a week anyway.

But after three days, watch for yeast floating up to the top of the brew. Skim it off before it sinks again. Save the skimmings for the next batch or to lend to other would-be brewers or bakers in need of live yeast. In a lidded, glass jar in a cool place, yeast will keep at least a month.

Beer made this way is pretty potent stuff. You can add corn to it to lighten it, or sugar to darken it. Other grains will make beer too, as I'm sure you know. Wheat beer is excellent, to my taste.

At any rate, I've given you only the bare bones of the process, and you should consult experts or other books, or be ready to employ a lot of trial and error. Failures are many, but faint heart ne'er won fair beer, I suppose. I take a dim view of the whole thing, remembering my father's beer. Bottles of it in our cellar had the peculiar habit of exploding occasionally. It was like living above a time bomb.

Malted barley, of course, is the soul of good Scotch whiskey. The smoky taste comes from drying the sprouted barley with burning peat. The leftover sprouts become the principal livestock feed on Scottish farms, at least in earlier times. Hundreds of little independent distilleries used to flourish in Scotland, making a profit by feeding the whiskey to people and the by-product, spent barley, to their livestock. Nothing wasted. Too bad America didn't follow that example in distilling whiskey. If it had, Appalachia today would be a prosperous place, not the land of poverty.

For barley soup or other foods, barley needs to be hulled. The blender will do a fair job if you then winnow or sift the hulls out. You can't get them all, just some. Since I have begun asking country people how to hull various grains, I've learned some rather novel methods. It seems that any hard striking action will work. As I have mentioned already, a combine with the beaters that strike the grain set overly close to the concaves will knock hulls loose on oats and barley. Some farmers tell me that if you blow the grain through a silage chopper against a silo wall, the considerable force of grain striking concrete will loosen about 60 percent of the hulls, be it barley or oats. Another method of hulling is to use a hammer mill, taking out the screen, and slowing down the RPMs to nearly half the grinding or milling speed. Roasting the grains before hulling greatly increases the efficiency of any method you use, as previously discussed in chapter 5.

Insects and Diseases

Like all the grains that have endured for so many centuries, barley is not a plant to roll over and play dead every time a disease or bug enemy comes along. Yellow dwarf virus may come close to destroying a field of barley occasionally if it attacks the plants at the seedling stage. Older plants under attack will be less stunted, but the top leaf will be yellowed and the grains in the lower parts of the heads blasted. An aphid carries the disease, but even insecticide applications have not helped control it. The aphid transmits the virus before it can be killed. Fortunately, the disease is not common.

Many fungal diseases bother barley as they do other cereal grains, especially in the humid South. In Georgia, spot blotch has been severe in the past, and farmers are advised to plant only the more resistant varieties and to resist heavy nitrogen fertilization.

Greenbugs are aphids that may attack barley if they can't find any wheat or sorghum. They've been around since 1882 at least, and have not overwhelmed us because of a host of predators, including lacewing flies, ladybugs, beetles, wasps, and syrphid flies. On occasion, the greenbugs still get out of hand and wreck some grain, but even from the viewpoint of a nonorganic commercial farmer, using an insecticide can do more harm than good.

Corn leaf aphids will feed on barley and other cereal grains in the South, curling leaf tips and turning them brown. I will quote what Arizona Extension researchers said about control back in the 1970s, so you don't think I'm just giving some nutty organic opinion of my own: "Infestations are greatest around the edges of the field. Thus, it is best to check throughout the field before deciding upon chemical control procedures. In most instances, lady beetles and other predators provide sufficient control. After the grain heads out, an infestation seldom warrants control."

Safe Storage

In storing barley, follow the same precautions as I gave for wheat. Though insect infestation problems with barley aren't as critical as with wheat, you still have to take care. The small amount you will want for your own use can be protected by heating or cold storage, as with wheat. For a small quantity, a metal barrel or two will provide enough room and will keep out rodents.

One reason I think the garden farmer should grow a variety of grains is because that makes controlling weevils and other storage insects easier. If you use up your barley from harvesttime in June until late summer, you aren't going to have bugs in it. They don't get a chance to become entrenched, so to speak. Then you can feed up your wheat and oats in the fall, then go to your corn and soybeans until the following summer, since weevils don't bother them much. If you have only a little harvested grain in bins for no longer than six months, and the grains are of more than one variety (so that one kind of weevil may not like one grain as well as another), and if you clean the bins or barrels out well when empty, you just aren't going to have a lot of insect damage.

In fact, on a well-planned, small-scale schedule, you could get by without having to store grain at all. Feed rye and barley right out of the field in early summer, wheat and oats in late summer, buckwheat and sorghum in fall, and corn out of the shock or off the stalk all winter. You'd have some loss from weather, but not much. Garden farmers have only begun to innovate commercial agriculture to their own purposes, and more new thinking like this will surely come along. We've all got a lot to learn. As an old English folk song that dates to the Middle Ages put it: "Neither you nor I nor anyone knows, How oats, peas, beans, and barley grows."

Barley Recipes

Highland Fling

2 pounds stewing lamb
½ cup split peas
1 cup barley
1 teaspoon marjoram
¾ cup chopped carrots
¾ cup diced potatoes
1 onion, chopped
Chopped fresh parsley and ground paprika to taste

- Cook 2 pounds stewing lamb until it is just tender, seasoning as you like it. Have the split peas and barley soaking in cold water during the last half hour or so of the meat cooking time. Add the split peas, barley, marjoram, and vegetables and cook until the vegetables are tender.
- The last thing before serving, salt to taste. Add chopped parsley and a sprinkle of paprika.

Yield: 6 to 8 servings

Hamburger Puff

1 pound ground beef
2 eggs, beaten
1½ cups cooked barley
1 small onion, grated fine
1 teaspoon salt

- Preheat oven to 350°F.
- Mix all ingredients, put in casserole dish, and bake about 40 minutes or until done.

Yield: 4 to 6 servings

Barley Pancakes

4 teaspoons dry yeast
½ cup lukewarm water
1–2 tablespoons honey
2 eggs
⅓ cup soy milk powder
¼ cup nonfat dry milk
1 cup water
1 cup barley flour
2 tablespoons oil
1 cup wheat germm

- Sprinkle the yeast over the surface of ½ cup warm water in a mixing bowl. Stir in honey and allow mixture to "work" in a warm place for about 25 minutes.
- Gradually blend in eggs. Combine the soy milk powder and nonfat dry milk with water, using a wire whisk, and add to mixture; then add the barley flour, oil, and wheat germ.
- Pour ¼ cup batter for each pancake onto a lightly greased griddle, over medium heat. When bubbles form on the surface, turn pancake and cook about 2 minutes longer, or until nicely browned on underside.

Yield: about 5 servings

CHAPTER EIGHT

Buckwheat and Millet

Buckwheat

People often learn what is good for them long before science figures out why. Buckwheat is an example. Traditionally, buckwheat was the all-American breakfast: the schoolboy's fortification against a snowy, two-mile walk to school, and the ax-man's fuel as he chopped his way across the frontier. The early American farmer, having stuffed himself with buckwheat cakes oozing maple syrup and surrounded by smoked sausage (how my mouth waters at the thought), could head for the barn with a whistle on his lips no matter how cold and cheerless the January morn.

I have grown buckwheat and eaten buckwheat cakes made from it, and, as good as they are, I have to wonder a little. Oats and wheat make cereals and pancakes just as tasty, and these grains yield better and more reliably. Why was the hardheaded early American so enamored of buckwheat? Why did he go right on eating it until "progress" made the hearty breakfast obsolete? And why, now, when eating breakfast is again fashionable, is buckwheat back in the limelight?

The Department of Agriculture has the answer. I quote from the monthly publication *Agricultural Research* of September 1974: "Agricultural Research Service's analysis indicates that buckwheat has an amino acid composition nutritionally superior to all cereals, including oats." Buckwheat's claim to such an honored status rests principally on its high content of lysine, a protein our bodies need but can't make and that is hard to come by in most other grains. That's why scientists are working so hard to perfect high-lysine corn.

So why isn't buckwheat a more commonly grown crop? In Pennsylvania and New York, where a considerable amount of buckwheat is still grown commercially, the grain is often called

Buckwheat

"the gambler's crop." Given a good market such as happened in the 1970s and seems to be happening again in the new century—when people who can't tolerate the gluten in wheat are eating breads and cereals made from gluten-free buckwheat—you may hit it big. Or you may strike out. Buckwheat is not a heavy yielder: 30 bushels per acre is considered a good harvest. The stems are brittle when mature and will break over in a hard wind and rain storm, leaving you with zilch. Buckwheat is naturally cross-pollinating and, until recently, could not be inbred because of self-incompatibility. Without inbred lines, better varieties were difficult to develop. Furthermore, the plant has indeterminate growth; in other words, the seed does not ripen all at the same time. When the first grains are mature, there will still be more green grains further up the stem, and flowering buds at the tip of the stems. If you thresh early, you will lose the seed just developing; if you thresh late, you may lose some of the early maturing seed, which can quickly shatter and fall to the ground after it is ripe.

But breeders are solving these problems. The Agricultural Research Service has developed tetraploid buckwheats that have more uniform seed size and thicker stems to resist lodging. These new varieties have not necessarily increased yield potential, however. Breeders have also discovered a buckwheat flower type that is self-compatible, and inbred lines are being developed, which in turn will lead to better varieties. Many garden-seed catalogs now carry common buckwheat varieties, and the Internet is full of sources.

The only farmer I know personally who has grown buckwheat commercially is one of the most artful and inventive persons ever to climb on a tractor, always coming up with new ideas. He pondered the fact that buckwheat could be planted late (midsummer) in the planting season and still make a crop. Having satisfied his curiosity on that account by experimentation and finding out in fact that late-planted buckwheat yielded better than buckwheat planted early, he thought about how he might use this characteristic to his advantage. His farm was already filled up with a cropping and livestock system he did not wish to alter. He went to his neighbors and asked them if he could rent their wheat fields after harvest. They readily agreed: it meant extra income for them and brought in more money than baling and selling off the wheat straw.

This genius of a farmer then baled the straw himself for the first return on his investment and then planted buckwheat. By now it was the middle of July. (He already had lined up a buyer willing to pay a good price for buckwheat.) In about seventy days, the crop was at its peak. To get around the problem of the buckwheat grains not ripening all at the same time, he cut and swathed the crop, which allowed most of the buckwheat "grains" (buckwheat is not technically a grain) to dry and mature more evenly. Then he combined it. The buckwheat straw he returned to the soil made up for the wheat straw he had taken off. Everybody won. He made more money, he said, than the landowner did on the wheat, and the owner had his land back for the following year's cropping.

Buckwheat has advantages that garden farmers and organic homesteaders might want to consider too. First of all, you can plant it just about anytime during the growing season and get some good from it. In earlier days, farmers pressed for time, as they usually were without today's big equipment, could plant buckwheat after the other pressing work was done, and that's usually when they did it. Today's garden farmers, spending perhaps forty hours a week at an outside job, can appreciate this advantage. If a series of rainy weekends gets you behind, and here it is July already and an acre or two left to plant, buckwheat is the answer. Or it can be planted after early vegetables or even after strawberries that you planned to tear up after the bearing season.

Buckwheat's main advantage for garden farmers, however, is that it makes an amazing green-manure crop. Andy Reinhart and Jan Dawson, who operate Jandy's market garden in Ohio, had an unbelievable crop of buckwheat this year. (I should not have been surprised—all their crops are unbelievable.) When I first saw it I was wary, however. It was so thick and tall I thought that it would be impossible to till it into the soil for green manure. I could at least understand why buckwheat is considered a good way to smother out weeds. You couldn't have raised a corn knife up through that mass of vegetation. As Andy showed me, though, the stems of buckwheat are rather fragile, and a heavy rotary tiller can chop through the dense growth for easy incorporation into the soil. And then the buckwheat will reseed itself for another crop.

Buckwheat has the ability to use phosphates in the soil that are unavailable to most other grains, according to the University of

Minnesota. Since buckwheat will grow in hard clay, it has the reputation of loosening such soils and making them more friable. But buckwheat roots do not grow deep in the soil, and the loosening effect probably comes more from the organic matter produced. As green manure, buckwheat will make *three* crops in one growing season.

Planting Buckwheat

Buckwheat has few disease or bug problems, which is another plus for organic growers. You can plant it by broadcasting the seed over a worked seedbed. No need to grow in rows for weed cultivation; in fact, you don't want to. Solid stands of buckwheat, as I already mentioned, will more than compete with most weeds. Simply rotary-till the soil, scatter the seed over the ground, and till lightly once more. A seeding rate of about 1½ bushels per acre is adequate.

In the garden I have grown buckwheat after early peas and gotten a good stand. One year, to test the claims of buckwheat devotees, I tried a patch on a hillside where the clay was leather-tough. In a very dry time in July, I planted the buckwheat there without fertilizer of any kind. With a little rain, the buckwheat grew luxuriantly. We harvested some, the birds ate some, and enough fell to the ground to give me another stand.

Harvesting

Buckwheat is best harvested with a combine, as mentioned, using the same adjustments and screens you use for oats. If you don't want to dry it in a windrow as described above, wait until after frost has killed the plants and the more mature seeds have had time to dry. This usually means harvesting at about 17 percent moisture, then drying the seeds down to the necessary 12 to 13 percent with artificial heat or spreading out the seeds very thinly in a dry environment.

Small amounts in the garden can be harvested by hand. Cut the stalks with a scythe (or sickle-bar mower), tie them into bundles, allow to dry well under cover, then proceed as with threshing wheat by hand. Buckwheat threshes easily. You can shake much of the seed out of the bundles when it is dry. Or rap each bundle over the edge of a bucket or the edge of a pickup truck bed. Or put the bundle in a sack and trample or flail as described earlier for wheat in chapter 3. Winnowing must then be done to separate out the chaff and stem bits.

With a garden patch of buckwheat, you can gather a cup or two at a time for breakfast from the standing plants, using your fingers to strip the dark brown, pyramid-shaped grains off the stems below the still-blooming tops of the plants.

Chickens like buckwheat. Rabbits do too. I just feed them the plants, with the grain still intact on them. A crop will not go to waste in any event, because, if you let the unharvested plants stand through winter, the wild birds will have a feast.

Buckwheat honey is most delicious. The nectar is especially beneficial for the bees because it is still available in the fall when they have to hunt much harder for flowers.

Processing buckwheat for table use is not so simple a task. The buckwheat grain, which looks like a tiny beechnut (from which word, *beech*, the *buck* of the word *buckwheat* derives, by the way) is mostly hull. The flour inside is nearly pure white. The ground-up hulls are good fiber, but like oat hulls, too many means less tastefulness. I like whole-buckwheat pancakes, but I prefer to have most of the hulls removed. With a commercial huller this is no problem, but at home, using a blender or kitchen mill, hulling is more difficult. We have used our blender to grind all grains (it will wear out sooner, however) and have found that if the buckwheat is toasted a wee bit or at least heat-dried well before grinding, the hulls will shatter off better, and many of them can be sifted out in a flour sifter. Well worth the trouble. Get some real maple syrup and some good homemade sausage to go with your buckwheat cakes. Instead of eating this breakfast when you first arise in the morning, go outside and work awhile first. Then you've set the stage for a truly great adventure in eating.

Buckwheat hulls make a neat mulch, but one that's expensive to buy. Unless you grow a large field of buckwheat, you won't get enough hulls to do you any good for mulch. But there's always a chance that some imaginative grower might see an opportunity in going seriously into the business of putting buckwheat cakes on every breakfast table, or at least most breakfast tables in his locality. If that person is you, don't forget to figure in the extra income from selling the hulls.

Let's muse on that awhile. A bushel of buckwheat weighs 48 pounds, or about 30 pounds of flour and 18 pounds of hulls when separated. A hundred acres of buckwheat producing a conservative 25 bushels to the acre equals 2,500 bushels, or approximately 75,000 pounds of flour and 45,000 pounds of hulls. Now multiply those figures by the price of a pound of buckwheat flour in your local grocery store and the price per 50-pound bag of buckwheat hulls at your local garden store. Need I say more? Of course, you'd need a commercial huller and a mill. But the possibility of profiting from the investment is there.

In case you think I've run out of nice things to say about buckwheat, here's one more. Buckwheat is a good source of rutin, a substance with medical value in the treatment of certain types of hemorrhaging. A ton of buckwheat plants makes one million rutin tablets.

Buckwheat Recipes

Buckwheat-Sesame Bread

¼ cup molasses
1 cup lukewarm water
2 teaspoons dry yeast
2 cups buckwheat flour
2 cups rye flour
2 cups whole wheat flour
1 cup sesame seeds
¼ cup oil
1 cup water

- Mix ¼ cup molasses and 1 cup lukewarm water. Add yeast.
- Let soak a few minutes.
- Combine the flours and sesame seeds in a bowl.
- Add ¼ cup oil and 1 cup water, blending well. Add the molasses-yeast mixture and work it into the dough with your hands. It will be sticky.

- Form dough into a ball. Place in an oiled bowl, then turn the dough over, so its entire surface is coated with oil. Cover bowl with a damp cloth and let the dough rise in a warm place for 3 hours, until double in bulk.
- Knead dough and form into 2 round loaves on cookie sheets. Let rise an additional 45 to 60 minutes. Meanwhile, preheat oven to 350°F.
- Bake in oven for 40 minutes or until done.

Yield: 2 round loaves

Buckwheat Blini (Pancakes)

¼ cup nonfat dry milk or soy milk powder
1 cup water
2 teaspoons dry yeast
4 egg yolks
1 teaspoon honey
4 tablespoons oil
1½ cups buckwheat flour, sifted
4 egg whites

- Combine nonfat dry milk or soy milk powder with water, using a wire whisk. Heat over medium heat until bubbles form on sides of saucepan. Remove from heat and cool to lukewarm. Add yeast and stir until softened.
- In mixing bowl, beat egg yolks until thick. Blend the yeast mixture into beaten yolks. Stir in honey and oil.
- Sift buckwheat flour and gradually blend it into batter, mixing thoroughly.
- Set bowl over a pan of warm water, cover, and let rise until double in bulk, about 1¼ hours.
- Beat egg whites until soft peaks form when beater is raised. Fold gently but thoroughly into batter.
- Preheat a lightly oiled griddle, over medium heat, until it is hot. Using 1 tablespoon of batter for small pancakes, 2 tablespoons batter for medium pancakes, and 3 tablespoons batter for larger pancakes, cook on the griddle until bubbles form on the edge and the pancake is golden

brown; turn pancake and bake 2 minutes longer. If pancakes begin to stick to the griddle, oil it lightly again.
Yield: 8 to 12 pancakes

Buckwheat Groats (Kasha) and Mushrooms

1 egg
1 cup buckwheat groats (ground medium)
2 tablespoons oil
1–2 cups sliced mushrooms
2 cups chicken broth
1 teaspoon salt

Beat egg well and add groats. Mix this very well to coat all the grains. Brown mixture in oil in a heavy skillet. Add sliced mushrooms. In a saucepan, bring the chicken broth to a boil and add to the skillet mixture together with the salt. Mix well and cook very slowly, covered, until all liquid is absorbed and the kasha is nice and fluffy, adding a little more broth if necessary.
Yield: 4 servings

Proso Millet

Millet is grown in the United States mostly for pasture and hay. Only proso millet is grown seriously for grain. It is used for animal feed, flour for humans, and birdseed mixtures. It is nutritionally superior to many of our common grains, containing more essential amino acids than wheat, oats, rice, barley, and rye. It lacks only lysine, the amino acid buckwheat is high in, making buckwheat and millet a good combination in your diet. Also, while most grains form acids in your stomach, millet, with its high alkaline mineral content, counteracts acids and is more easily digested. Millet, not rice, is the basic carbohydrate food in China, especially northern China. The Hunzas, whose reputation for health and longevity is well known, eat millet regularly.

The word *millet* is used to refer to plants in four different fami-

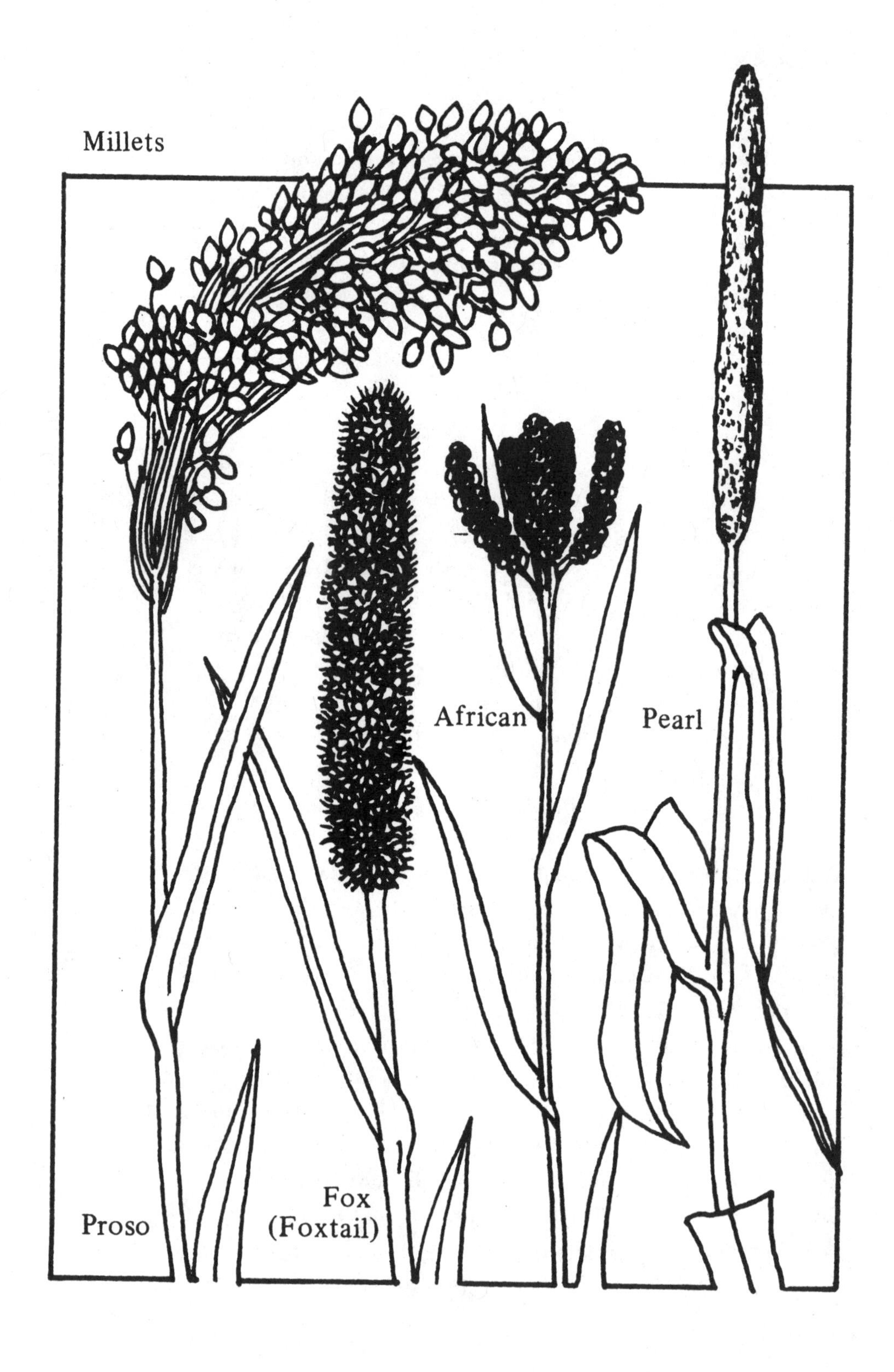
Millets
African
Pearl
Proso
Fox
(Foxtail)

lies, and therefore leads to a tremendous amount of confusion, including mine. Sellers of field seed in the United States talk about Japanese, German, Hungarian, African, common, proso, pearl, browntop, foxtail, and variations thereof. And these terms do not necessarily refer to the same plant in different parts of the country, either. So, armed with my ever-trusty *Taylor's Encyclopedia of Gardening, Horticulture, and Landscape Design*, 4th ed., (Riverside Press, 1961) and supported by innumerable phone calls to seedsmen throughout the United States, I shall attempt to identify all the millets and colloquial names thereof. But, mind you, I won't claim infallibility for my categorization. One man's colloquialism is another man's slang.

There are three different families of millets and a fourth kind so-called, which is not really millet at all. Let's dispatch with this fourth one first. If you are in Texas or surrounding states, you can buy and grow what is called "African millet." This plant is really a sorghum, a tall form of kafir corn with a proper name of *Sorghum unlgare* var. *caffrorum.* It is grown for pasture and/or hay, though not extensively. African millet might also be referred to as mock-orange cane, orange sorghum, or even sumac, in Texas.

Proso millet, *Panicum miliaceum,* is the only millet grown for food in the United States. It is sometimes called broomcorn millet because the open heads of the plant resemble small broomcorn heads. That differentiates it from the foxtail- and cattail-shaped heads of other millets. So far as I know, no millet is sold under the name "broomcorn millet" anymore, but interest in proso is increasing as a food grain for the driest parts of the country. This family of millets is the one used from earliest times for grain and flour, especially in India, China, Japan, Manchuria, and Russia. Proso is usually milled for livestock as well as humans because the seed coating is so hard. Chickens can handle it whole. The seeds are about the size of peppercorns, and either red, yellow, or white in color. Newer varieties are white.

"Foxtail" millets, *Setaria italica,* are grown for emergency hay, silage, and pasture, especially where weather is invariably dry. They make better forage than proso because they are finer-stemmed and not hairy-stemmed. What we in this part of the country call barnyard millet is a foxtail millet that volunteers in midsummer and grows better than other grasses in drought conditions. Once it heads out, which it does rather quickly, the livestock don't much like it.

So-called browntop and Japanese varieties are not always included with the foxtail millets. That's where scientific classification and colloquial names really get confusing. So-called Japanese millet grows in the North; browntop in the South. Both may grow to a height of 5 feet and will regrow after cutting, yielding more tonnage of forage than any other millet. These millets are finer-leaved and finer-stemmed than sorghum-Sudan grass and so dry faster for haymaking. Browntop is used in the South for cover and feed for quail and other game birds, but a Georgia farmer I once talked to said he considered it good hay for dairy cows, too.

The third group of millets are the pearl millets, botanical name *Pennisetum glaucum*, or *P. americanum*. Pearl millet, rather than resembling a small foxtail grass, looks more like a cattail reed head, and is in fact, often called cattail millet in the Southeast. Pearl millet is grown almost totally in the South. It threshes free from the hulls, which might make it more desirable for the garden farmer processing millet for table use. But it is not as desirable for grain or flour as proso.

Millet can be planted either by broadcasting or by drill seeding, at the rate of about 35 pounds per acre for all kinds of millet—a little more if broadcast. Don't worry about getting it a little deep, as it will come up from 4 inches to 5 inches down, I'm told. But 1 inch to 3 inches is a better planting depth. Millet, like buckwheat, can be planted late in the season. In fact, it is often grown as an emergency crop after another grain crop has failed.

Millet won't compete, pound for pound, as hay or pasture with legumes in nutritional value, but it has other advantages. Thirty days after you plant it you can be using it. No legume can make that claim. It has good insect resistance and is relatively free of disease, which together with its ability to grow on rather poor land makes it a desirable crop for any organic grower not wanting to use commercial fertilizers. As hay it can produce several tons per acre in three months, and as seed, 50 bushels per acre or more. It's a good emergency forage when things go wrong and you need forage fast. And, of course, the same would be true of the grain. Chickens will do well on it; just toss them the whole stalks and let them peck out the seeds, or let them free-range through the standing millet. And researchers, particularly in the drier parts of the Great Plains, keep up ongoing programs at the land-grant colleges on millet improvement. As one of them told me: "If we ever run out of irrigation water, millet could become a very valuable food source, as it is in other dry regions of the world."

Nutritionists point out that proso millet is highly adaptable to various recipes, as it has an almost bland taste, with just a slightly nutty flavor. They say that it can easily be used by itself or in combination with other grains in casseroles, breads, stews, soufflés, stuffings, cereals, or eaten plain with butter, gravy, or a vegetable sauce. Brown it first in a skillet with a small amount of oil, then use it as you would any other grain. Browning enhances the nutty flavor.

Millet Recipes

A coworker and one time *Organic Living* editor at Rodale Press, Ray Wolf, shared this idea with me. "You can prepare a cereal that will serve a dual purpose by adding one part millet, one part sesame seed meal, and five parts water to a baking casserole or a double boiler and cooking it until done, about 45 minutes. This can be eaten as a cereal, or allowed to cool and congeal, at which point it can be sliced or prepared as patties, which can be used as a type of (cornmeal) mush. It can be reheated and served with cheese melted on top."

Millet Bread

⅔ cup millet flour
⅓ cup barley flour
1 cup grated raw carrots
1 tablespoon honey
1 teaspoon salt
2 tablespoons oil
¾ cup boiling water
3 egg yolks
3 egg whites

- Preheat oven to 350°F.
- Combine millet flour, barley flour, grated carrots, honey, salt, and oil in a bowl; mix well. Gradually add the boiling water, mixing thoroughly.
- Beat egg yolks until light and lemon-colored. Stir into the flour and carrot mixture.
- Beat egg whites until they peak, but are not dry. Fold carefully into batter.
- Oil an 8 × 8-inch pan or a 9 × 5 × 3-inch loaf pan. Line bottom with brown paper and oil again. Spoon mixture into prepared pan and bake for 30 to 40 minutes in preheated oven or until done.
- Remove from oven: set pan on wire rack and allow to cool for about 5 to 10 minutes. Remove from pan. Serve warm.

Yield: 5 to 6 servings

Cashew-Millet Casserole

2 cups water
1 teaspoon salt
½ cup millet meal (whole millet that has been ground in blender)
3 tablespoons oil
1 medium-sized onion, chopped
1 cup unsalted raw cashew nuts (ground in electric blender, ½ cup at a time)
3 eggs, slightly beaten
⅔ cup wheat germ
½ cup nonfat dry milk
¼ cup chopped parsley
¼ cup chopped pimento
½ cup water
⅛ teaspoon of the following dried herbs: ground mace, sage, rosemary, ground marjoram

- Preheat oven to 325°F.
- Prepare millet meal: In a saucepan, bring 2 cups of water to a boil; add salt and millet meal to rapidly boiling water very

slowly, stirring constantly with a wire whisk to avoid lumping. Place in top of double boiler and continue to cook 15 to 20 minutes, or until millet has absorbed all water. Remove from heat and cool slightly.

- Heat oil in skillet and sauté onion.
- Combine cooked millet and ground cashew nuts in mixing bowl. Add slightly beaten eggs, wheat germ, nonfat dry milk, sautéed onion, chopped parsley, and pimento. Add ½ cup water, blending together thoroughly. Stir in herbs. Adjust seasoning according to taste.
- Turn mixture into an oiled 1½-quart casserole and bake in preheated oven, uncovered, for 45 minutes or until firm and lightly browned. Serve immediately.

Yield: 6 to 8 servings

Millet Soufflé

4 cups boiling water
½ teaspoon salt
2 tablespoons oil
1 cup millet meal (whole millet that has been ground in blender)
4 egg yolks
¼ cup nonfat dry milk
1 cup water
½ teaspoon crushed dill seeds (optional)
3 tablespoons minced chives or grated onion
1 cup grated or shredded sharp cheese
4 egg whites

- Preheat oven to 350°F.
- In saucepan, bring 4 cups water to a boil; add salt and oil. Gradually add millet meal to boiling water, stirring constantly with a wire whisk until all of it has been incorporated. Place in top of double boiler and cook for about 20 to 30 minutes, or until all water has been absorbed. Stir mixture occasionally. Remove from heat and cool slightly.
- In large mixing bowl, beat egg yolks until thick. Combine

nonfat dry milk with water, using a wire whisk, and beat gradually into yolk mixture.

- With beater set at medium speed, blend cooked millet into yolk mixture until thoroughly combined. Stir in dill seed, chives, or grated onion. Add grated cheese.
- Beat egg whites until soft peaks form; gently fold into millet mixture.
- Pour mixture into an ovenproof casserole (2-quart size) and bake in a preheated oven for 35 to 45 minutes. Remove from oven and serve immediately.

Yield: 6 servings

CHAPTER NINE

Rice

The Oldest Garden Grain

In the Western world, rice is hardly a major grain, at least not compared to corn or wheat, either commercially or for home production. However, rice is, worldwide, the oldest and most successful garden grain of all, and affords ample proof that the production of a major grain almost entirely from small garden farms is valid and practical. In Asia, rice has been a garden crop for at least four thousand years. The typical Asian farmer may grow a hectare or two at the most, and millions—literally millions—of families are able to live a fairly comfortable life because of rice plots smaller than that. One could say without exaggeration that the culture of most of Japan, China, India, and Southeast Asia is built on—and survives because of—a cottage rice industry. We Americans may not possess the keen Asian taste for rice, or may live where rice cannot be grown, but we can learn from rice the economies of grain gardens and how to develop a technology that serves such economies rather than a technology that forces grain production into the hands of a few human land hogs, some of whom already tell me they would rather not be called farmers anymore. (I have honored their request.)

Agriculturists might argue about which grain, rice or wheat, is the most important in the world at the present time. Certainly more people eat rice than eat wheat, but more wheat is consumed. Rice is not nearly as "commercial" as wheat. The bulk of the former is produced at home for home use. In fact, the United States, which produces only about one percent of the world's rice, is the leading exporter of the grain! A comparison between rice growing in America and Japan can be almost soul-shattering. A father-son team in Texas may handle 500 acres of rice or more, but barely make a good living by our standards. That many acres of rice in Japan supported one hundred families comfortably up until a couple of decades ago, and still does to some degree. And yet we insist that we are the efficient ones.

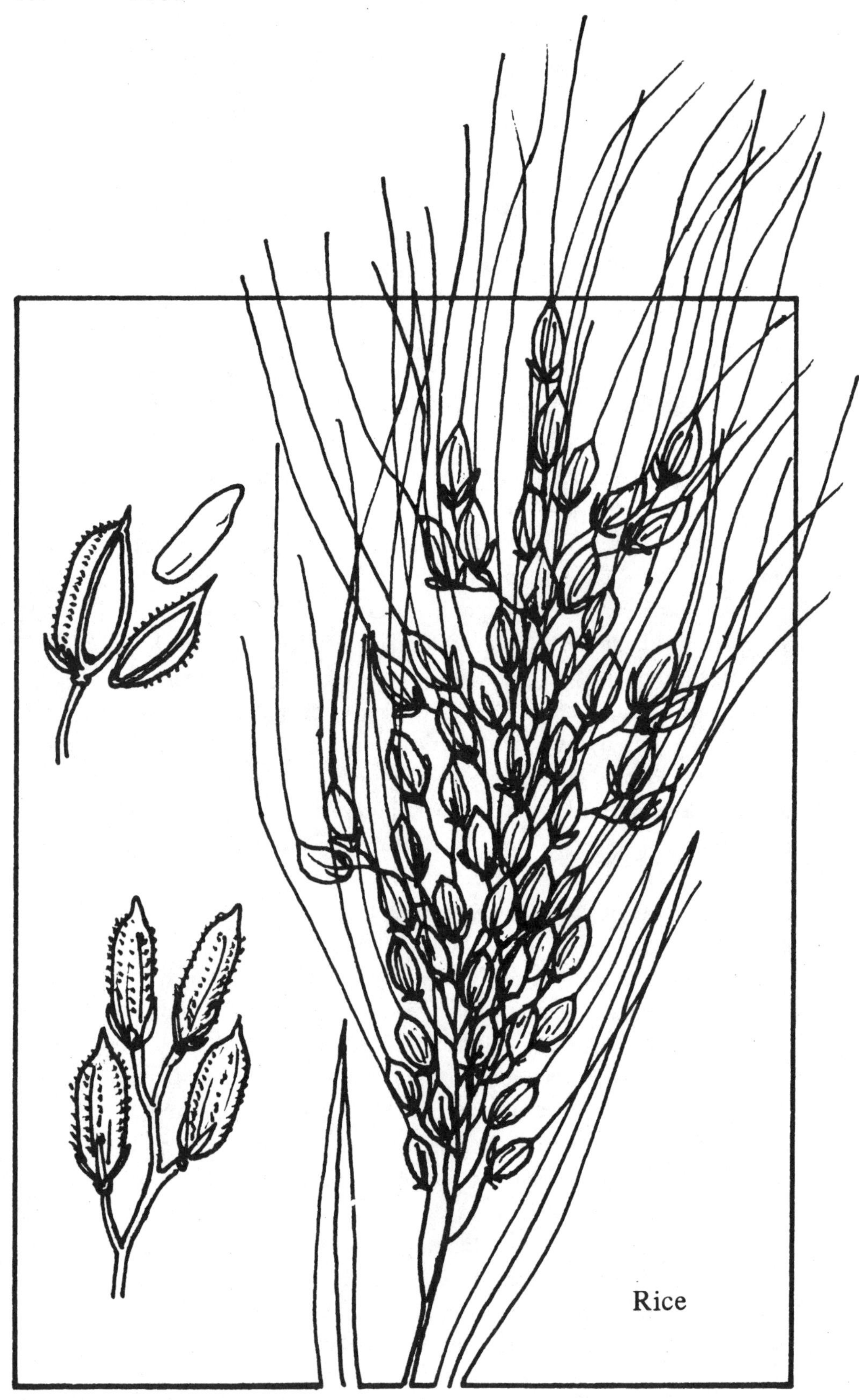

Rice

The Culture of Rice

Whether grown with Western machinery or Eastern backbone, rice is not the easiest grain to produce. It prefers a long growing season and warm humid weather. It is grown profitably in our country only in the Southwest, mostly in Arkansas, Louisiana, Texas, and California. However, rice could probably be grown farther north, at least for home use. The Japanese have learned how to grow it successfully as far north as Hokkaido, which has a climate similar to our southern New England. Upland rice—varieties that grow without flood irrigation—will produce a crop in Thailand at 4,500 feet above sea level, and at twice that elevation in the Himalayas.

Upland rice does not yield as well as irrigated rice, though improved varieties show more promise. Upland rice is grown about like spring-planted wheat. Where level land or marshland is available for wet rice production, upland rice is usually discouraged.

Wet rice, or irrigated rice, can be direct-seeded in the paddy or field or started in a bed and transplanted. With direct hand-seeding, the number of hours to raise a crop can be reduced by about two-thirds. Much of the seeding in the United States is done by airplane.

The typical Asian farmer has been loath to switch to direct seeding despite the labor-saving advantages. He seems to prefer longer working hours to the higher cost of chemicals and machinery that would be necessary in place of labor. Agricultural experts seem to think his attitude is stupid. But the Asian farmer knows he is not going to gain a whole lot in net profits anyway by adapting new technology, and he runs the risk of becoming much more vulnerable to financial disaster when he substitutes cash and chemicals for labor. His American counterpart hasn't learned that yet, but, oh my, the lesson is well underway. When technology offered the American farmer the bait, he swallowed. Technology said: "All you farmers are farming 500 acres and barely making it. I can make it possible for you to farm 1,000 acres and get rich." The farmers accepted this as faith, not understanding that for every 500-acre farmer who went to 1,000 acres, some other farmer had to give up his 500 and go to work at something else. Now, when the 1,000-acre farmers find they aren't doing a whole lot better than when they farmed 500 acres, the technological answer is to farm 2,000 acres. And again, the farmers believe it.

Farming as Art

I suppose the Japanese farmer grumbles about the backbreaking aspect of his hand methods sometimes, but like true garden farmers the world over, he seems to like it, too. At least there is ample evidence that he derives satisfaction from it. He makes an *art* of his agriculture, as anyone who has seen well-tended, terraced rice paddies on a Japanese hillside can appreciate. It *is* art. Kusum Naizi, in *The Lonely Furrow: Farming in the United States, Japan, and India* (University of Michigan Press, 1969), pointed out that, while the Japanese farmer's yard may be weedy, "never his fields. Yet he knows full well that those last blades of grass that he pulls out of his paddy so laboriously do not affect his production by an ounce. The transplanting of rice seedlings is done with similar excessive care. It is neat, precise, and meticulous, like an embroidery on silk."

Agriculture for art's sake is not limited to the Asian farmer. Ronald Blythe in his excellent portrait of English rural life, *Akensfield* (Penguin Press, 1969), emphasized the same propensity for art at the expense of economics among English farmers before industrialization. They too would row, hoe, cut, bind, and rank in precise patterns and spacings, so that fields took on the order and geometry of a formal garden.

The same sensitivity to art in agriculture marked the American farmer before industrialization. Farmsteads were built with an eye to beauty as well as utility. The farmstead scenes that find their way into picture frames today are all derived from an era before high technology engulfed farming. Today few farmers build art into their workaday world. They can't afford to if they did want to. Old implements were always decorated with painted swirls, stencils, engravings. That took time, and now technology has decreed that time is a commodity to be priced, bought, and sold on the same callous basis that slaves were priced, bought, and sold. Technology has taken away the farmer's time from him, and time is the most precious possession a man owns. So today the banks are built of marble and exhibit a taste for art; the barns, if built at all, are simply large, hideous collectors of manure from overcrowded animals.

My close neighbors, now gone, would have understood the Japanese farmer very well. They were of the old school of farming. I've told about Uncle Carl, who walked the rows of his soybean field, cutting weeds with his hoe. He spurned the herbicides he could have used because he knew they would not keep his fields nearly as clean as his hoe would. He cultivated with his tractor, too, but it would not get all the weeds either. Like the Japanese rice farmer, he was not satisfied until he removed *all* the weeds. His motives had nothing to do with profit, which a few weeds would not hurt. He did not have to make a profit anyhow. He had been a good and successful farmer all his life. He had all the money he desired. He hoed the weeds because he wanted his field to look beautiful. And to farmer eyes, his weedless field *was* art. In all this county full of herbicides and monstrous cultivators, only his field was without weeds. And farmers stopped along the road to admire it, and admire the work that made it so. He admired it too. He shared with "old-fashioned" farmers a wisdom the new technologists can't comprehend. He had raised his daily work to the level of art, while the technologist slaved away all his days hoping to reserve a little time in the end for art purchased from an antiques store. Whose "economies of scale" were the wiser?

While the experts preach the advantages of direct seeding to him, the traditional Japanese farmer goes about making his nursery bed where he first "roughs out" the work of art that will also be his livelihood. He presses the seeds into the ground carefully. They must not be completely covered with soil, though he might cover them lightly with mulch. The bed is then flooded. When the primary leaves emerge, the bed is drained. Only very gradually, though, so the tender sprouts are not too quickly exposed to air and direct sunlight. More floodings and irrigations may follow, depending upon temperature, moisture, weeds, diseases, insects, the alternate irrigation and drainage serving to avoid too much of one problem while making sure there is enough of a countering solution. The process is scientific, but its application to the realities of *this* nursery bed, in *this* year—that is pure art. In about forty days, the young rice seedlings are transplanted into the paddy at precise spacings; close enough together for a maximum number of plants, but still allowing enough space for a man to walk while weeding. Again the field is flooded and drained alternately on a schedule based on the needs of that climate and that soil and that particular season. When ripe, the grain is cut by hand, the sheaves bound, dried, and then threshed using small hand- or motor-operated threshers. Output per man, by our standards, is extremely low, but efficiency in terms of number of people fed per unit of fossil energy used is extremely high. As F. H. King pointed out in his classic work, *Farmers of Forty Centuries,* in 1907 Chinese, Korean, and Japanese garden farms were feeding five hundred million people, almost twice the population of the United States today, on an area smaller than all the improved farmlands in the United States at that time. And doing it without any of today's big machinery, commercial fertilizers, or herbicides.

Growing rice by the hundreds of acres, as in the United States, is a very different and expensive operation. Seed and chemicals are most often sprayed by plane now, followed by a light irrigation. Then, after the plants are about six to eight inches tall, standing water is kept in the fields until the crop begins to ripen. The very large fields laid out with levees on the contour are a marvelous feat of engineering, and a kind of art, too.

Scientists are now trying to cut down the use of some chemicals in commercial rice production, not only to cut expenses but also because the chemicals sometimes become nearly useless. For instance, because the fields stand in water for a good portion of the summer, rice land is a haven for mosquitoes. But bombarding them with insecticides has resulted in immune mosquitoes. Newer, more biological controls are now being tested, with some success. In Arkansas, mosquito fish stocked in rice fields at rates over one hundred fish per acre gave "very good to excellent control" after seventy-two hours. Insect-growth regulators also are showing promise in controlling the rice-field mosquito.

Weeds are difficult to control in rice fields without chemicals, but simplistic herbicide applications are no longer thought to be the solution. Researchers are working with a fungus that attacks northern jointvetch, a serious weed in rice. If successful, the venture would be a real breakthrough in biological weed control.

In the United States, three kinds of rice are grown: long-,

medium-, and short-grained rice. Long and medium are mostly Southern rices; short-grained rice is grown largely in California. As with any grain, many varieties exist of each type. If you want to try a small plot of rice as described at the end of this chapter, use a variety that's recommended for your area. If none are, proceed at your own risk.

Rice from the field has a tight hull on it like oats and must be hulled. Rolling or abrasion easily removes the hull. Once hulled, rice is processed as brown rice, white rice, or parboiled rice. Humans historically have preferred refined white rice for eating, which is unfortunate, since white rice is the least nutritious of all. Brown rice is largely unrefined and therefore contains almost all the nutritious bran. White rice is polished rice, that is, the bran has all been abraded away. The supposedly "practical" reason the bran is taken out of rice (or wheat, or whatever) is that the bran has a high oil content, and, if milled into the flour, soon causes rancidity. In other words, milled brown rice won't keep very long and should be milled only as needed. Parboiled rice is a kind of compromise between brown and white. The rice is cooked and dried before polishing away most of the bran. The cooking drives some nutrients, particularly B vitamins, on into the grain, thus preserving them in the flour even though the bran is removed.

White rice is about 90 to 94 percent starch and 6 to 10 percent protein. In removing the bran, 85 percent of the oil, 10 percent of the protein, 80 percent of the thiamine, 70 percent of the minerals and crude fiber, 50 percent of the riboflavin, and 65 percent of the niacin are lost, according to the USDA. A comparison of vitamin content of the three kinds of rice tells the story (see table 5).

TABLE 5. VITAMIN CONTENT OF RICE

	Thiamine	Riboflavin	Niacin
White	0.14	0.096	3.2
Brown	0.785	0.082	8.185
Parboiled	0.419	0.084	9.606

Quantities given are milligrams per cup.
Source: USDA Nutrient Data Laboratory online database at www.nal.usda.gov/fnic/foodcomp/search/.

Growing Rice Organically

Rice is being grown organically on a commercial basis, and profitably when demand for brown rice is good enough to support a high market price. During the early days of the brown rice boom around 1970, Wehah Farms in California turned over its large commercial rice operation to organic methods. As reported in *Organic Gardening and Farming* magazine in 1971, Wehah found that rice *could* be grown organically, though yields were less than those produced normally with chemicals. Instead of using pesticides to control tadpole shrimp, Wehah followed a carefully timed schedule of raising and lowering water levels in the fields to overcome, at least in part, the difficulties caused by this pest. Good sanitation and close attention to intervals between irrigations helped control seedling diseases, rice water weevil, and rice leaf miner. Crop rotation controlled weeds to some extent, and fish stocked in the fields kept the mosquitoes at bay. Instead of burning old rice straw, Wehah returned it to the soil. Chicken manure and green-manure crops were plowed into the soil for more fertility.

While few commercial rice growers would consider operating that way, the small homesteader and gardener can achieve success with rice using organic methods. Since I have no hands-on experience growing rice in northern Ohio, I was fortunate enough to find a remarkably detailed article by David Spiekerman and Junsei Yamazaki, in *Organic Gardening and Farming* from December 1975 (p. 64 ff.), giving an excellent step-by-step account that anyone in the proper climate should be able to follow successfully. Part of this account follows:

> The technique we used to grow our rice is basically the method Japanese farmers have used for centuries. We used no chemicals, poisons, or machines to grow our rice. My partner, Junsei Yamazaki, grew up among rice paddies in Japan, and showed me traditional, organic, hand methods from start to finish. The whole effort was simple and smooth. Any healthy person can do it.
>
> The plot of land we chose for our paddy had two assets. First, the soil had been under grass for countless years, and it was clean and naturally fertile. Second, the location

was right next to our water well, so we conveniently used a short, rubber hose to pump the water into the paddy. However, the best water for irrigating rice is stream water because it contains more oxygen and minerals than well water.

Two disadvantages of this location did not appear to us until after we had planted. First, the paddy was situated on the north side of a row of large, leafy almond trees and therefore did not receive full sunlight until the afternoon. Secondly, we had no space to dig a small, warming basin for the cold well water. Ideally, the water entering the rice paddy should be warm to promote maximum growth and health of the rice plants. Our water emerged cold from the well and traveled a mere four feet to the paddy. Consequently, our rice grew slower than usual because we did not give it full sunlight nor warm water all day.

Our rice paddy was 10 × 18 feet. Because our ground was level, we shaped the paddy into a rectangle. On uneven, sloping ground, it is advisable to contour the shape of the paddy as the ground dictates. We created the paddy a few days before transplanting by an ingenious technique. First, we turned over the soil with a spade, working it loose and friable to a depth of one foot. The topsoil was a deep, clay loam in excellent health. Around the perimeter of the 10 × 18-foot plot, using a special spade with a long, narrow head, we dug a trench one-and-a-half feet deep and six inches wide. The soil dug out was placed along the tops of the four outer sides of the trench to form four dikes, each one foot high.

Next, taking sheets of black plastic, we placed them against the outer walls of the trenches and over the dikes. The plastic retarded seepage of water from the paddy into the bone-dry ground surrounding it. Then we leveled the soil between the trenches, which lowered the paddy bottom. We used a long, flat board and our eyes to level the paddy bottom. It was essential for the paddy bottom to be flat so that water would stand at an equal depth over the paddy. We tested our paddy by flooding it before planting. . . .

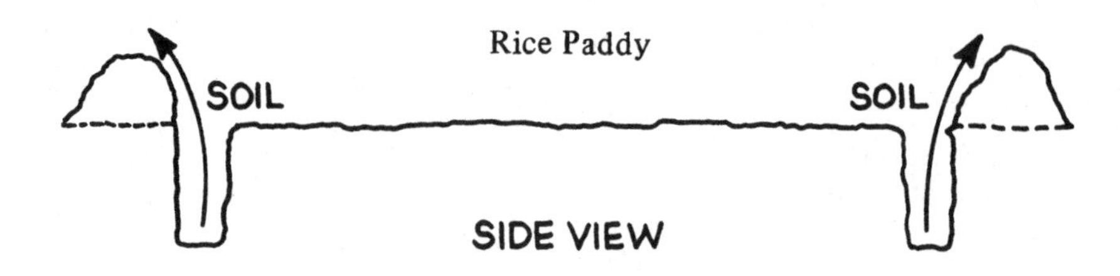

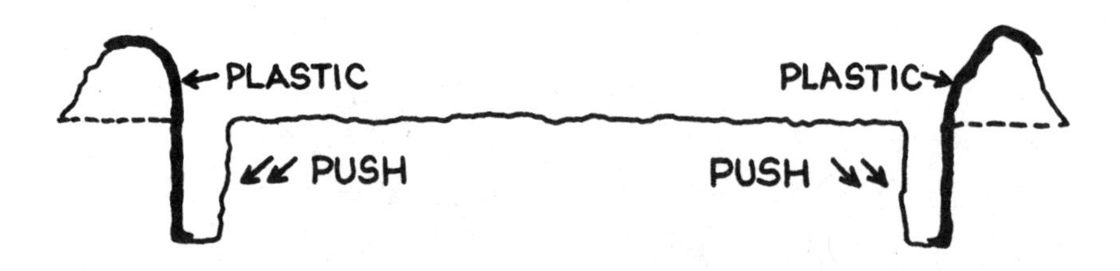

We used seed Yamazaki had brought from Japan years ago. Taking two pounds of seed, placing it in a bucket, adding cool water three-quarters of the way to the top of the bucket, and then adding a few tablespoons of unrefined sea salt, we swirled the seed, water, and salt around and around with our hands. When we stopped swirling, the viable, strong seeds sank to the bottom, while the weak seeds floated to the top. We removed the weak seeds by hand and swirled again, repeating the process until no more seeds floated on top of the water. Then we poured the strong seeds into a wire strainer and washed the seeds with cool water to clean away the salt. . . .

The day after selecting viable seed, March 28, we placed the seed in a small, cotton bag and tied it closed with string. Placing the bag full of seed in a bucket, we set a three-pound stone on top of the bag. Then we filled the bucket to the brim with water. Keeping a hose in the bucket, we ran just enough fresh water in the pail to keep a very slight dribble of water falling down the outside of the bucket. This amount of overflow was an indication that water was circulating through the bucket at a slow, gentle rate. We checked the overflow a few times daily. The seed spent 23 days soaking outdoors in the bucket. Rice can soak in water for up to 30 days this way. The water was cool and never warm. This long soaking starts the germinating process and indicates the unique strength of rice. Soaking also kills weak seed, which means that only strong seed will be planted in the ground.

On the 20th of April, we took the bag out and opened it. Each strong seed had a tiny, white sprout at one end. For one day, the seed lay in the bag on top of a stump out in the sunlight and air, which dried it slightly. The following day we planted the soaked seed, but not in the paddy; our technique required transplanting. So we prepared a small, seedling bed two feet by four feet in a sunny, warm spot of soil. The bed was slightly raised above the surrounding ground, which made it easier for us to remove the young rice plants for transplanting. Adding rich leaf compost, sand, and water to the bed, we planted the sprouted seeds by broadcasting them over the ground. The seeding density was thick enough to cover the ground almost completely.

Next we covered the seed lightly with sand. It is important not to cover the seeds with too much sand, which can cake and crust, making it difficult for the sprouts to break through to sunlight. On top of the entire bed, we laid a thin layer of rice straw and tied string across the straw lengthwise to prevent the wind from blowing the seeds and straw away.

It may be necessary to screen the bed from birds that like to eat the tender rice sprouts.

Every two days, we watered the seedlings by hand. Sometimes it rained, but rain is not common or dependable in late April and May in Chico. One week after planting the seeds, tiny green shoots appeared. We weeded every day to allow the rice maximum opportunity to utilize the soil nutrients. A seedling bed is a very fertile, concentrated plot of soil. The quality of the rice harvested is influenced by the soil condition of the seedling bed. Our young rice plants grew for 35 days in this bed to a height of four to five inches. About 40 days is ideal for developing strong rice plants suitable for transplanting.

On May 26th on a hot afternoon, we transplanted into our rice paddy. Wetting the soil to a muddy condition, we removed the seedlings from their bed by hand, being careful not to damage their roots. We carried them over to the paddy. There we had marked off 14 rows lengthwise, eight inches apart. Planting the seedlings in rows allowed us room to walk and to hand-weed during the growing season. Planting by airplane or by broadcasting provides no open space to walk and weed. Commercial rice growers must resort to poisonous weedicides sprayed by airplane. Rice paddies attract a wide variety of weeds, which grow faster than rice. It is best to pull the weeds out of the soil and then plunge the top of the weed plant back into the mud where it will decompose and feed the rice.

Three of us bent over with a handful of seedlings and firmly pushed them into the mud, two or three plants in clumps six inches apart down each row. It took us 20 minutes to complete the transplanting. It would take two healthy people one day to transplant an acre of rice. For the first 20 days after transplanting, we kept water at a depth of one inch in the paddy all day long. Our pump steadily fed water into the paddy at a slow rate. We turned the water off at night, and the water in the paddy seeped slowly into the depths of the soil. After one hour, no water would be standing in the paddy. Freeing the soil of water during the night allowed more oxygen to enter the root zones of the rice plants. Incidentally, no mosquitoes could breed in the paddy without a continuous, standing body of water. In the morning, it took roughly one hour to fill the empty paddy to a one-inch depth with the pump turned on fully.

The seedlings took one week to establish their roots in the wet mud and to stand straight. Occasionally, a seedling would lose its hold and float on the water but would not die. We simply picked it up and pushed it down in the mud again. The commercial rice growers flood their paddies to depths of six inches or more. Our technique flooded the paddy to a depth of no more than one inch of water. The slow circulation of oxygen-rich water to the roots of the rice plant is the essential point to irrigating rice. Maintaining a depth of one inch of water allows this to happen.

After 20 days passed, we watered only in the morning because we wanted the warm, afternoon sun to hit the soil directly around each rice plant. The sun's rays stimulated new rice shoots to grow from the base of each plant. The fecundity of this seed was amazing to watch. Each plant was capable of producing 12 new stalks in a season, all capable of bearing heads under optimum conditions.

From the middle of June to the 1st of August, the rice plants grew nearly two feet in height. New shoots continued to appear. During this time, a Great Dane chewed about one foot off the tops of 30 plants on his stroll through the paddy. The injured rice was strong enough to recover the lost growth in a week and assume the same height as the rest of the rice. Frogs were living and playing in the paddy at night. Grasshoppers, dragonflies, butterflies, and other small insects frolicked among the deep green leaves. Our cats slept in the cool, wet paddy during the hot afternoons. Once a gopher dug a hole in the paddy, and the water leaked out. We placed two mothballs down the hole and filled it with soil. The gopher never bothered us again.

Only twice in the five months the rice was in the ground did we weed. (Commercial growers use a formidable arsenal of chemical poisons to eliminate weeds and pests; next to cotton, rice receives the heaviest dose of chemicals in American agriculture.) Our experience never suggested such a need. The paddy was a thriving ecosystem. Our role as farmers consisted primarily of regulating the water. The paddy took care of itself.

During the first week of August, plants at the sunnier, north end of the paddy began to head. Keeping the paddy covered all day with one inch of water is critical during the heading stage. Excessive rain and wind, which we did not have at that time, can retard the flowering and self-pollination of rice. The heading moved from the north end to the south slowly because of the shade of the almond trees. All the plants headed by the second week of September. Some plants were barren, others were part fertile, part barren. Seeds which appeared white were either insect damaged or barren. These may have been five percent of the total seed production.

The rice started to yellow by mid-September, and the heavy, maturing heads bent downward. One week later on September 21st, we stopped flooding the paddy. We permitted the rice to dry for a month in the paddy, two weeks longer than the necessary two weeks. On October 19th, we eagerly harvested the golden grain. Cutting with sickles and bundling it by hand took us one hour. Then we threshed the rice for another hour.

Threshing was easy. We built a wooden frame two feet tall with a metal grate on top. Beating a bundle of rice against the grate a dozen times released the seed from the stalk. Placing the seed on a plastic sheet, we cleaned it by throwing it up in the windy air. For one week, we dried the rice in the sun.

Milling the rice to eat is a real chore. The husk adheres firmly to the grain and offers excellent protection for it. The commercial mills crack about 20 percent of the grain in the husking process. A hand machine sold in Japan is available to the small rice farmer. The most primitive method of removing the husk is to place the grain in a stump with a depression carved out of it and then hammer the rice with a wooden mallet. The force of hitting rubs the grain against the stump and scratches and loosens the husk. This breaks a lot of the grain and does require you to separate the grain from the husk with a fan or with the wind. It is best to mill the rice in small quantities as it stores well in its husk and removing the husk begins the slow process of nutrient loss. . . .

Our yield was 30 pounds of paddy rice. We planted

> approximately 1,550 seedlings in the paddy, which amounted to three ounces of seed rice. Our paddy was roughly 1/215 of an acre, so to get an idea of how much rice one acre will produce using our method, multiply 30 times 215. The yield would be 6,450 pounds per acre from 40 pounds of seed. The commercial, organic rice growers get a yield of 2,500 pounds per acre. Chemical growers of rice get 5,000 pounds per acre but must plant 100 to 200 pounds of seed per acre. The rice straw we cut from the paddy has multiple uses. It is a valuable feedstuff for livestock. It makes excellent mulch. It can be woven into matting. In comparison with other growers, we did well. But our true pleasure was assisting the growth of a truly remarkable plant. The enjoyment we got watching our rice grow cannot be measured.

Never having tried to grow rice myself in Ohio, I feel a little uneasy talking about it. I wanted to give a very detailed account of it in this book, though, as much to satisfy the reader's curiosity as anything, but I bet that not many garden farmers have either the will or the time to grow rice in the meticulous manner described above. Too tedious. I doubt that one would have to attend to such minute details and would find shortcuts that would make growing rice much more straightforward. I also have a hunch that growing upland rice is easier still.

In recent years there has developed, at least in California, another source of revenue from rice. The straw, baled, is in demand for building/construction uses. Also, if progress continues to be made in turning plant fibers into fuel, there should be more demand for straw of all kinds to make ethanol or, more likely, biobutanol, which appears to be a more efficient fuel to process from grain straw.

Rice Recipes

Below are a few recipes. Your own favorite cookbooks will have many more.

Greek Lemon Soup (Avgolemono)

6 cups chicken stock
½ cup raw, brown rice
Salt to taste
1 whole egg
2 egg yolks
¼ cup lemon juice
2 tablespoons parsley, freshly snipped
⅛ teaspoon ground cayenne
Freshly chopped dill to garnish

- Put chicken broth into a heavy saucepan and bring to a boil. Add rice to soup, season with salt, and cook until rice is tender, 15 to 20 minutes.
- Put whole egg and two egg yolks into a medium-sized bowl; beat with a rotary beater or wire whisk until light and frothy. Slowly add the lemon juice, beating together thoroughly.
- Just before serving: dilute the egg-lemon mixture with 1 cup hot broth, beating constantly with a wire whisk until well blended. Gradually add the diluted mixture to the remaining hot soup, stirring constantly. Bring almost to the boiling point, but *do not boil or the soup will curdle.* Stir in the parsley and cayenne; adjust seasoning.
- Remove from the heat and serve immediately, garnished with freshly chopped dill.

Yield: approximately 6 cups

Cheese Quiche in Brown Rice Shell

1½ cups cooked brown rice
3 eggs
¼ cup nonfat dry milk
1 cup water
¼ teaspoon salt
Dash of freshly ground black pepper
Dash of ground nutmeg
1 tablespoon whole wheat flour

1¼ cups shredded sharp cheddar cheese (or part natural Swiss and part cheddar)

- Preheat oven to 375°F.
- Press cooked rice into an oiled, 9-inch pie plate. Bake in oven just until dry (about 5 minutes); cool.
- Using a medium-sized bowl, beat the eggs until light and fluffy. Combine nonfat dry milk and water with a wire whisk and add to eggs, along with salt, pepper, and nutmeg.
- Add whole wheat flour to shredded cheese; toss lightly but thoroughly. Put into cooled rice shell, spreading to edges of crust. Pour egg-milk mixture over all. Place in oven and bake 10 minutes. Reduce heat to 325°F and continue to bake 25 to 30 minutes longer, or until filling puffs up and is golden brown.
- Remove from oven; allow to set for about 5 minutes; cut in wedges and serve.

Yield: 6 servings

Sprouted Lentils, Bean, and Rice Salad

½ pound pinto beans or kidney beans
1 pound fresh green beans, cooked
2 cups cooked brown rice
1 cup diced celery
½ green pepper, diced
¼ cup chopped pimento
¼ cup sprouted lentils
½ cup oil
½ cup wine vinegar
1 tablespoon honey
1 teaspoon salt
1 teaspoon pepper
1 medium-sized red onion for garnish, sliced thin

- Soak pinto or kidney beans overnight in water to cover. Do not drain. Using soaking water, cook the beans until just tender. Don't overcook. Drain, reserving the cooking liquid for soup.

- Combine green beans and pinto or kidney beans, rice, celery, green pepper, pimento, and lentil sprouts.
- Combine oil, vinegar, honey, and seasonings. Toss salad in dressing with the onion rings.

Yield: 10 servings

Brazil Nut Cookies

½ cup soy flour
2 cups brown rice flour
½ teaspoon salt
½ cup nonfat dry milk
1 cup oil
10 tablespoons honey
2 eggs, slightly beaten
1 teaspoon pure vanilla extract
1½ cups ground or finely chopped Brazil nuts

- Preheat oven to 400° F.
- Sift flours, salt, and nonfat dry milk together.
- In a large mixing bowl, combine oil, honey, slightly beaten eggs, and vanilla extract. Using an electric beater set at medium speed, beat mixture until ingredients are thoroughly blended.
- Gradually beat in sifted flours and nonfat dry milk. Stir in Brazil nuts. Refrigerate dough for 1 to 2 hours.
- Take rounded teaspoonfuls of cookie dough and roll into 1-inch balls. Place 2 inches apart on an oiled cookie sheet; flatten with a glass. Bake in oven for 8 to 10 minutes or until golden brown around the edges. Remove from oven and place cookies on a wire rack to cool.

Yield: about 5 dozen

CHAPTER TEN

Some Uncommon Grains, Old and New

Wild Rice

Unless you live in northern Minnesota or certain parts of the upper Great Lakes and Canada, wild rice probably means little more to you than something gourmets eat or that you order at a restaurant when you feel rich. And, most likely, you leave the table wondering why anyone would pay that kind of money to eat the stuff regularly. You didn't really think it was that tasty, right? An Ojibwe Native American would say you just didn't know how to fix it.

Wild rice is ill-named. It is not rice, and it is becoming less wild every year. Wild rice belongs to a different genus of grasses (*Zizania*) than rice (*Oryza*) and resembles the latter more because of the way it grows in water than in appearance. Wild rice grows much taller (up to 9 feet) than real rice, and its leaves are much wider, about 2 inches compared to a ½ inch for rice.

Moreover, wild rice is in the process of being domesticated into a commercial farm crop, what the agribusiness community proudly calls another example of modern technology making a once esoteric, costly food available to every consumer. With apologies to the Wild Rice Institute, I think the domestication of wild rice is absurd. Rapacious man has found one more enclave of nature to exploit for profit, but this time the endeavor may be self-defeating.

Domestication of wild rice first involves making fields out of marshes, fields that can be flooded and drained at will. Goodbye marsh. Getting enough water then to flood the paddy becomes a problem. So much water is needed that growers often have to obtain permits to use it, and some have been charged with violating water-use laws and antipollution regulations.

To raise yields from about 50 pounds per acre (in a natural stand) to a profitable 500 to 700 pounds per acre means a large infusion of chemical fertilizers. To control leaf spot, the paddies have to be burned or sprayed. Rice worm is controlled by spraying insecticides. Weeds must be controlled, too. Birds, deer, muskrat, and other wildlife will harm the crop. Get out of the way, nature. Humans must make the North safe for domesticated wild rice, that terribly essential food for a hungry world.

Then the rice is harvested by big combines. But wild rice shatters very badly and too much is lost. Solution? Why, new strains, of course. These have been provided, and more are on the way. So we will have a wild rice that is not wild, with a taste that doubtless will not please the Ojibwe and Chippewa and the gourmets who buy it now.

By the rice growers' own admission, there is not now a real market for expanded production of wild rice, but the growers assume that the consumer can be persuaded to buy it. If production expands, dragging the market behind it, the price will fall.

Already some growers fear that big increases in production could destroy the market. In the meantime, the American Indians and the working-class whites who have traditionally depended on the natural wild rice harvest for extra or essential cash will see their business go to pot. Who wins?

Real Wild Rice

Natural wild rice production is a fascinating, even romantic story, a frail life system held together in modern times by a colossal program of government protection. For centuries Native Americans could harvest wild rice using traditional methods that insured the continuity of the rice crop and a goodly supply for the migratory waterfowl that depend heavily on the grain. When the white man took over, he generously left the Indians with the sole right to harvest wild rice.

Like all his other promises, the white man soon broke this one. There was a little money to be made in wild rice, and it was unfair that the Indians should get it all, he said. American business decided that this was a "monopoly." Can you imagine a blacker pot than that calling a kettle black? The law was changed so that everyone could harvest wild rice except on Indian reservations and certain other areas left to the Indians. I would bet that, if the price of wild rice should climb very high, the Indians would lose those exceptions too.

The rules and regulations that wild rice harvesters must abide by are unbelievable. But perhaps fortunately so, because, man being the greedy creature that he is, the wild rice would soon disappear without heavy protection. In years when the price is high, a group of wild rice poachers can glean $100 [in 1975 dollars] from one night's work.

Harvesting time is set by regulation, usually from about 10:00 A.M. to 2:00 P.M. every other day during a season set so that migratory fowl are insured enough of the grain on their journey south. Game wardens and sheriff's deputies mobilize like a small army all up and down the upper regions of the Mississippi: boats, cars, planes, and helicopters roar up and down the rice stands keeping everyone honest. The cost of the protection is probably higher than the worth of the crop.

But anyhow, at the sound of a siren, harvest begins. It must be done exactly as the Indians did it who knows how long ago—only two people to a canoe or flat-bottomed rowboat. One must pole the boat along while the other harvests. The harvester uses two wooden sticks, one in each hand. With one stick, he or she pulls a bunch of wild rice stalks to the boat, bending the heads over the gunwale. With the other stick, he or she strikes the heads a swift blow, shattering the seed into the bottom of the boat. It's hard work. And hardly efficient. But enough seed falls into the water for next year's crop. And when the price is over $1 a pound, everybody makes a little money. And at least the crop is not inundated with chemicals, nor is the ecology of the marshes destroyed as in the commercial paddies.

According to my sources, wild rice can be found from the mouth of the St. Lawrence River to central Manitoba, south to Kansas and Virginia, and even around the coast to Louisiana. But the eastern variety is somewhat different and is not harvested. No wild rice will grow in seawater. If you live where it does grow, you can establish plantings that will draw birds better than almost any other plant. Wild animals as well: waterfowl, songbirds, upland game birds, deer, moose, muskrat. Maybe you can even grow some of this nutritious, high-protein, truly organic grain for yourself. But if you buy it in a gourmet shop, just remember: just because it's "wild" it ain't necessarily so. It could come from commercial, chemicalized paddies unless it definitely says organic. And even then, well, I'm the overly suspicious type. If you want to try the genuine article, harvested and processed traditionally, search the Internet for the Native Harvest website; the color and flavor of truly wild rice (the Ojibwe call it *manoomin*, or "the good grain") is far different from most commercially grown wild rice.

I tried to establish wild rice in my farm pond and in the creek that runs through our place. I tracked down seed from the University of Minnesota as I remember. Didn't have the Internet in those days. My efforts to grow it were all in vain, however, so as I briefly try to tell you how to do it, remember that I failed.

Take special care of the seed. If it dries out, it will not germinate. Commercial growers store it all winter in bags in water at a temperature of about 35°F. Better to harvest the seed from one plot and carry it immediately to the new marsh or backwater and toss it in.

The plant grows best in water from 6 to 12 inches deep, but will tolerate water depths up to about 3 feet. In summer, before the growing plants break above the surface, they can turn a river or marsh to a beautiful lacy green color. By September, the tall plants hide the water almost completely and turn yellow with maturity.

In some years, bad weather hurts the wild rice crop seriously. Then harvesting is prohibited so that there would be enough seed for next year. Wildlife authorities say the wild grain will prevail, if man exercises caution.

Wild Rice Recipes

Wild Rice and Squash Soup

½ cup uncooked wild rice (approximately 1½ cups cooked)
2 tablespoons oil
1 medium-sized summer squash or zucchini
1 green onion, including tops
½ clove garlic
1 tablespoon chopped parsley
¼ cup chopped mushrooms
⅛ teaspoon thyme
⅛ teaspoon basil
4½ cups chicken stock

- Sauté rice in oil until all grains are well coated. Chop all vegetables very fine or put through a food processor or blender. Add vegetables to rice and continue to sauté for a few minutes. Add seasonings and then 1½ cups of the chicken stock and bring to a boil.
- Cover, lower heat and simmer for 40 minutes. Add the additional 3 cups chicken stock, bring to a boil and serve.

Yield: 4 cups

Golden Herbed Wild Rice

1½ cup uncooked wild rice (approximately 4½ cups cooked)
¼ cup butter

2 medium-sized onions, finely chopped
1 shallot, finely chopped (optional)
½ teaspoon dried basil
½ teaspoon dried tarragon
½ teaspoon dried chives
Pinch of dried thyme
1 tablespoon chopped parsley
3 cups chicken stock
1 tablespoon tomato paste
½ teaspoon ground turmeric

- Sauté rice in butter until all grains are well coated. Add onions, shallot, and herbs and continue to sauté for a few minutes. Add stock, tomato paste, and turmeric, bring to a boil, cover and simmer for about 25 minutes, or until rice is tender and all liquid is absorbed.
- Mold the rice in buttered custard cups or a fluted mold. Cover mold and keep warm until serving time. Unmold onto platter around meat or in the center, or serve separately.

Yield: 4 to 6 servings

Note: This recipe can also be used as a stuffing for fowl or as a filling for steamed summer or winter squash.

Triticale

Wild rice may be our newest domesticated grain, but triticale is our newest grain without qualification of any kind. It is also the first grain crop made by man rather than by natural processes.

Triticale is a cross between wheat and rye, the name deriving from the Latin for wheat, *triticum,* and for rye, *secale.* Triticale is not a new idea. For a long time plant breeders have tried to cross the two plants to team the productivity of wheat with the hardiness of rye. But though thousands of crosses, even millions, were made, the offspring were infertile or of such poor quality as to be valueless. Finally in 1967, in the midst of a full-scale program of triticale breeding initiated between the International Maize and Wheat Improvement Center in Mexico and the University of Manitoba,

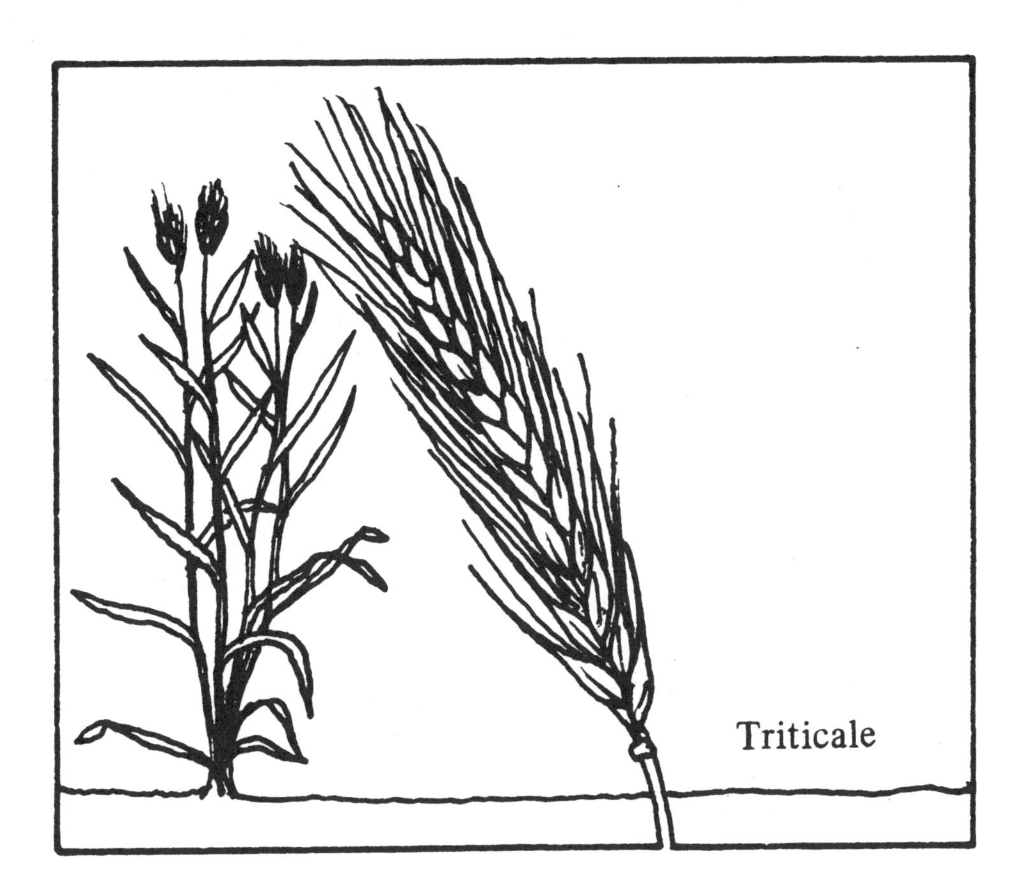

a chance pollination occurred in a research plot that was fortunately observed by researchers. From that catch, the modern and sometimes sensational triticale crosses eventually were developed.

In the late 1960s triticale became the darling of agricultural journalists. They could call the grain everything from "miracle crop" to "solution to world hunger" and get away with it. It was high in protein, and, compared to most grains, productive where wheat was not; it could be pastured, contributed more nutrition to bread, and was not hard to grow. I remember how excited we all were on the staff of *Farm Journal* magazine. We procured a few seeds, which one of the editors planted in his backyard. We watched them grow through the summer with the kind of wonderment one feels when he first walks up to a 747 jet.

Alas, as is often the case with new varieties, triticale has not always lived up to its pregame statistics. Near Bowling Green, Ohio, I have seen excellent triticale grown by natural-farming methods that yielded 55 bushels per acre, nearly 5 feet tall, and without lodging. The crop was sold at quite a good price to specialty markets. But at the same time I have seen poor stands not three counties away. The data from college test plots comparing triticale and wheat invariably indicate much higher yields from wheat. While livestock feeding tests show the value of triticale—it can be used to replace nearly half the corn in feed rations—the yields of improved wheats are high enough to outscore triticale in total protein per acre.

In the East, Southeast, North, Northwest, and particularly Northeast, triticale still suffers occasionally from low fertility. Not enough of its flowers fruit. This condition not only results in low yields, but it also contributes to the development of ergot, a very toxic fungus. Farmers I know who have grown triticale say it winter-kills more easily than wheat too.

When asked, plant breeders across the Corn Belt will point out these present disadvantages, but will then proceed to sing the praises of triticale for some future time. Maybe so. But improvements in wheat, oats, and barley may continue to outdistance triticale development. If you want to try a row or two for fun in the garden, then you can judge for yourself whether it will be worth a bigger crop. Especially for your own family's diet, triticale will give you a better mix of amino acids used along with wheat. Grow the grain exactly as you would grow wheat.

Seed is available for gardeners from garden-seed catalogs. If you try to make bread from it, be aware that, like rye, triticale contains little gluten. You'll have to mix in some wheat flour. Better varieties, or some more suitable to your area, should be available from local farm-seed salesmen. At least they can steer you in the right direction.

Triticale Recipes

Triticale Nut Drops

2 eggs, beaten
½ cup oil
½ cup honey
½ teaspoon vanilla
2½ cups triticale flour
½ teaspoon salt

1 teaspoon cinnamon
¼ teaspoon crushed anise seeds
½ cup chopped walnuts
Butter for greasing cookie sheet

- Preheat oven to 375°F.
- Combine eggs, oil, honey, and vanilla. Combine flour, salt, spices, and nuts and add to liquid ingredients. Drop by teaspoonfuls onto a greased cookie sheet, press flat with the bottom of a glass that has been dipped in water, and top with a walnut piece if desired. Bake in oven for 12 to 15 minutes or until golden brown. Cool on rack.

Yield: 4 dozen

Triticale Egg Bread

2 tablespoons honey
½ cup lukewarm water
2 tablespoons dry yeast
2 teaspoons vinegar
1 cup milk
2 teaspoons salt
2 tablespoons oil
2 cups triticale flour
4 cups whole wheat flour

- Dissolve honey in lukewarm water. Sprinkle yeast over surface and set aside for 5 minutes to activate. Add vinegar to milk and heat just to lukewarm temperature, stirring constantly, until the milk curdles. Remove from heat and pour into large bowl.
- Add salt, oil, eggs, and triticale flour to soured milk.
- Add yeast mixture, stirring in well, then add the whole wheat flour, reserving about 1 cup for kneading.
- Turn dough out onto a well-floured board or counter and knead for a full 5 minutes.
- Place dough in an oiled bowl, turning it over to oil the entire surface. Cover and put in a warm place to rise for about 1¼ hours or until double in bulk.

- Punch down dough. Let rise again for ½ hour or so.
- Form into 1 large loaf and 1 small loaf, place in buttered bread pans and let rise for another ½ hour, or until dough is slightly rounded over the top of the pan. Preheat oven to 350°F. and bake for 30 to 35 minutes or until done. Remove loaves from pans and cool on rack.

Yield: 1 large and 1 small loaf

No-Knead Triticale Bread

4 teaspoons dry yeast
⅔ cup lukewarm water
2 teaspoons honey
3 cups whole wheat flour
2 cups triticale flour
3 tablespoons molasses, unsulphured
⅔ cup lukewarm water
½ tablespoon salt
⅓ cup wheat germ
1⅓ cups lukewarm water
½ tablespoon butter
1 tablespoon hulled sesame seeds

- Sprinkle yeast over lukewarm water. Add 2 teaspoons honey. Leave to "work" while preparing the dough.
- Warm flours by placing them in a 250°F oven for about 20 minutes.
- Combine yeast mixture with molasses mixture. Stir this into the warmed flour, then add the salt and wheat germ and finally the 1⅓ cups lukewarm water. The dough will be sticky.
- Butter a loaf pan (9¼ × 5¼ × 2¾ inch), taking care to grease the corners of the pan well. Turn the dough into the pan. No kneading is necessary. Smooth dough in pan with a spatula that has been held under cold water to prevent stickiness. Sprinkle sesame seeds over top of loaf. Let it rise to the top of the pan in a warm, draft-free place. Meanwhile preheat oven to 400°F.
- Bake in oven for 30 to 40 minutes, or until the crust is

brown and the sides of loaf are firm and crusty. Set pan on rack to cool for about 10 minutes, then remove loaf from pan and cool completely on rack before slicing.

Yield: 1 loaf.

Spelt

Among organic growers, there has been a renewed interest in spelt, one of the oldest grains grown commercially. As far as I can figure out (because no one will exactly agree with me), organic farmers wanted a grain easy enough to keep separated from wheat in the marketplace. Only a few farmers were growing spelt, so making it an organic grain might take away some of the temptation that a wheat farmer might have to try to sell chemical wheat at the higher organic price. At any rate, two close acquaintances of mine, both organic farmers, tried spelt. One of them quit, but the other is still enthusiastic about it.

Spelt is cultivated like wheat, and is similar to it. Unlike wheat, it will not thresh out cleanly. Each kernel is surrounded by a tight glume that won't shatter loose easily. But livestock eat it just fine, glume and all. Spelt is generally taller than wheat and therefore lodges more easily when grown on good soil. Heads are long, like rye. Planting rate is about 2½ bushels per acre.

One advantage of spelt is that people who can't eat high-gluten wheat can tolerate spelt *sometimes.* Specialty bakeries offer spelt bread. Find one in your area by checking on the Internet.

If spelt isn't old enough to attract your curiosity, contemplate the ancient grain, emmer. Used to be you could find emmer in a very large dictionary, but hardly anyplace else. But the increase in interest in all garden grains has brought a sort of comeback for emmer. The Land Institute near Salina, Kansas, grows emmer as a possible grain, along with wheatgrass, from which they are hoping to develop a perennial wheat, one of the ongoing goals of the institute. Emmer is an ancestor of wheat.

There are other esoteric grains like kamut, a close relative of wheat, that are getting new attention. If you look up Eden Foods or Mary Jane's Farm on the Internet, to name but two, you will find an amazing assortment of foods from an amazing assortment of grains. Semolina wheat is the hard white durum wheat that is typically used to make pasta, and there's farina, which is made from soft durum wheat, and couscous, the national food of Morocco—it just boggles the mind the number of grain foods now available. Eden Foods sells a special gourmet spaghetti made from kamut, for example. Needless to say, you can grow your own pasta from many of the special wheat-type grains, or buy them from mail-order catalogs like Mary Jane's. In the interest of full disclosure, I know Mary Jane personally and wrote about her when she was first starting out in selling organic foods by mail. Her business is now an amazing success story. Although she grew up in commercial farming, she had the vision to see beyond it, to the possibilities in something seemingly as quaint as garden grains, and then make those possibilities come true.

I should probably include a couple of paragraphs about right now on how to make your own pasta from scratch. It is easy enough to do. But you can find the instructions all over the place and little kitchen-sized pasta machines are easily available. We've made lots of pasta and dried it on those folding indoor clothes racks. We think the superior taste is well worth the effort.

Farro

Farro is closely related to spelt. Experts argue about how closely related the two plants are, but farro's Latin name is *Tritium dicoccum,* and it is a very ancient grain, one that is becoming popular again in fancy European restaurants and among a growing number of gourmet fans in the United States. The Internet is full of references. Sometimes spelt can be used in place of farro, but most gourmets disagree and advise that when you buy the grain, make sure it is *T. dicoccum.*

Quinoa

Pronounced "KEEN-wa," this ancient South American grain (*Chenopodium quinoa*) is a member of the goosefoot family and is technically not a real grain, but referred to as a pseudo-grain. It has become quite popular in recent years among gourmet chefs and natural foods customers. Quinoa is extremely nutritious, whether cooked like a grain or sprouted. In fact, it is the only sprout that contains all the amino acids essential for humans.

Like many other Andean crops, quinoa is sensitive to day length and cannot always be grown successfully in our latitude. However, breeding and selection work is ongoing. Seeds of Change sells quinoa seed and says that the plants bloom in late summer to produce richly colored head spikes in a range of burgundy-orange, yellow, white, and pink shades. It might be worth trying a small garden planting at home, for the plant's ornamental qualities, if nothing else.

Quinoa seed contains saponin, a soapy, unpleasant-tasting substance, and so the general recommendation is to rinse seeds in a couple changes of cold water before cooking. The taste is nutty and delicious and useful for salads, stews, or pilaf.

Flax

Like quinoa, flax is not technically a cereal grain, though its protein-rich seeds are used in livestock feed, are added to bird feed, and would be good human food, as the ancient Greeks well understood. But whether grain or not, flax deserves some mention in any book written for garden farmers interested in self-subsistence. For two centuries at least, flax, from which linen is made, was one of the main sources of fabric for American clothing. With so much interest in homespun wool and cotton, we are only a step away from a renewed enthusiasm for homemade linen and linsey-woolsey. Making linen from flax is not really that much more difficult than making your own cotton or wool yarn, and even if most of us will never try it, we need to be aware of its practical possibility.

Humans have grown flax for thousands of years. Mummies in Egyptian tombs are wrapped in fine linen, indeed a much finer linen than we know how to make now. Some climates and soils have proved historically more suitable for producing fine linen than others, Belgium being perhaps the most famous. Flax seems to do better in cool, dry climates and today is grown mostly in

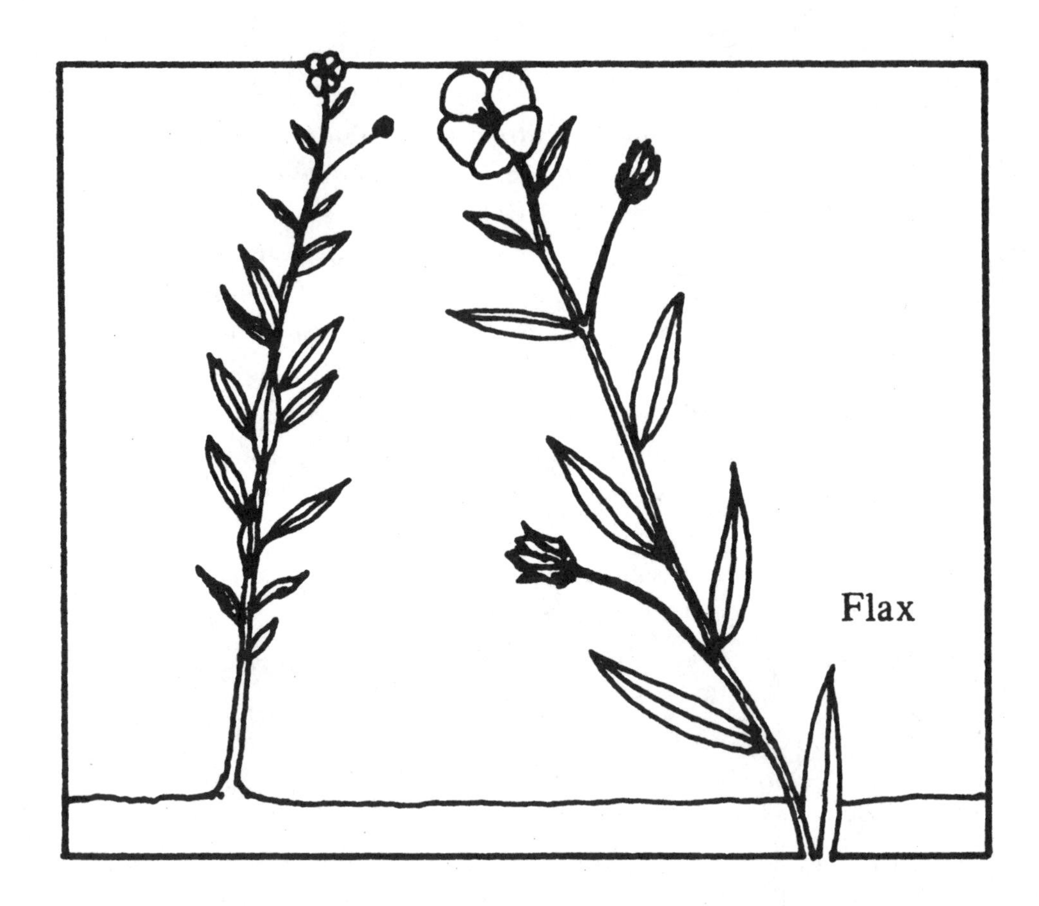

the more northerly areas: Canada, Russia, and our northernmost Plains states. Flax production has been on a decline in our country. Very little is grown for linen in the United States anymore, but instead for linseed oil. The fibers from the straw have been utilized for cigarette paper and other fine papers. Acreage in our northern states has been falling mostly because higher wheat prices have lured farmers into shifting land formerly used for flax into wheat. But the rise of water- and rubber-based paints, which has reduced the demand for linseed oil, has been the long-range cause of flax decline.

Both flaxseed and flaxseed oil are excellent sources of omega-3 fatty acids. There's a catch, though, especially to flaxseed oil, which pregnant women should not consume, as well as other people with specific physical conditions. Also, there's a danger that flaxseed oil might be confused with linseed oil, which it is, but industrial linseed oil is somewhat toxic and should never be consumed by anyone. A good Internet site to check on both the pros and cons is herbwisdom.com.

Flax *can* be grown in just about any state, though perhaps not commercially. Some varieties have flourished at different times in Oregon, California, Texas, even Georgia. Before 1900, Kentucky and Ohio were the leading producers. Wilt disease, against which earlier varieties had little resistance, was a common reason why acreage in flax kept shifting to new, untainted soil.

Flax is grown much like wheat, drilled (or broadcast) about an inch deep, at a rate of 2 to 3 pecks per acre, or even up to a bushel in humid climates. If you want to grow flax for the fiber, plant at the heavier rate, as thick stands discourage branching and therefore make longer, smoother stalks for longer fibers. Plant flax early in the planting season. Normally good soil should not need extra fertilizer for flax; nitrogen would only encourage weeds, which flax competes with poorly. Never plant flax in a weedy field, the old government bulletins say, and don't plant it after sorghum, millet, or Sudan grass. When the roots of these crops are rotting in the soil, they can cause injury to the growing flax.

Flax grows fairly fast, and by June in the North the pale blue blossoms begin to appear. Dr. C. D. Dybing, flax specialist at the University of South Dakota, who so kindly answered my questions about flax back in the 1970s, said that a field of it at bloom time looks like a beautiful lake eddying in the breeze. In Canada, he says, where rape, flax, and wheat are grown in rotation, you can see huge rolling fields of yellow rape, blue flax, and, at that time of year, gray-green wheat next to each other, and the scene is breathtakingly beautiful.

Flax keeps on blooming, somewhat like buckwheat does, but later blossoms do not develop mature seed, and so in the North there is no reason to wait for them to mature. When most of the seed is about ripe, the crop is cut and swathed, and ripening is completed in the swath. Then it is harvested with a combine. Average yield is about 10 bushels per acre, but can be as high as 25 to 30 bushels.

Making Linen

If you wanted to make linen, you would not cut the stalks, but would pull them so that you could obtain longer fibers. Pulling by hand was the age-old way, though mechanical pullers were developed by the end of the nineteenth century. You can expect at least a ton of straw per acre from which to extract fiber and usually more, perhaps twice as much. A tenth of an acre would give a hobbyist about all the fiber he or she would want to spin, at least during the learning years.

The Goshenhoppen historians at East Greenville, Pennsylvania, keep alive the craft of linen-making, and I was lucky enough, in the 1970s, to observe the whole process performed at their Goshenhoppen Folk Festival. You almost need to watch such a demonstration to learn the craft of linen making. The flax plants, which have been pulled, allowed to dry, and the seed heads cut or combed out, are soaked in water or left on the ground exposed to the weather for several weeks to rot the woody stems around the fibers. The rotting process is called *retting*. When the stems have sufficiently deteriorated, the plants are dried. Then the flax, a bunch at a time, is passed through a tool called a flaxbrake, which breaks the stems in several places. The flaxbrake is about the size of a sawhorse, which it somewhat resembles. The flax stems are laid across three (usually) fixed wooden boards or bars, and a set of other boards or bars, hinged at one end of the sawhorse affair, are lowered or dropped between the fixed lower boards. As the top boards or bars pass between the bottom, the flax stems are broken.

Next, the broken stems are scraped or "scutched" with a swingling knife. The worker holds a bunch of broken flax in one hand and scrapes or swingles with the knife in the other hand. The knife looks about like a corn knife, but is made entirely of wood. While swingling, the worker rests the bunch of stems on the sharpened edge of a perpendicular board that measures an inch or so thick and 6 or so inches across and stands about belt-high. The woody stem pieces surrounding the fibers fall to the ground under the onslaught of the swingling knife. Next the worker passes his handful of fibers through a hetchel or hacksel. The hetchel looks like a large brush, only with steel teeth (or combs) spaced widely apart. This step, called hackling, straightens the fibers and pulls out all remaining stem pieces caught in them. The processor has left in his hand what looks quite a bit like the end of a horse tail, only the "hairs" are much finer and fluffier. When you are watching the process, this appearance of a swatch of fibers at the end seems rather sudden and a bit magical. It takes about as long to describe the process as to do it.

The fiber then is ready to be spun into yarn, and quite nice yarn, too, that would make very durable clothing. Incidentally, craft historians tell me that flax can be spun on either the smaller "flax" wheel or the larger "wool" wheel. I use the quote marks because the experts are by no means agreed that the smaller, so-called flax wheels were used exclusively for flax. At any rate, once you have the flax fibers, spinning proceeds as with any other material.

If you are experimenting with flax, you might also like to know that the raw oil pressed from the seeds is much sought after by violin makers to finish the wood of their finest instruments.

If you don't live in flax country, getting a small amount of seed may be difficult for you. Distributors of birdseed might be able to get it for you. Some garden catalogs handle flax seed, but mostly the ornamental kind. Again, the Internet to the rescue.

Cotton, Sunflowers, and Safflower

Having talked about flax, I feel obliged to mention the obvious, that cotton in the South provides, or can provide, a garden farmer the same advantages that flax does in the North, including the excellent protein value of cottonseed and cottonseed cake. And having said that, you should also be aware of the protein and oil value of field-grown sunflowers and more esoteric oil seeds like safflower. But I have to draw the line somewhere, so I will draw it along the boundary between the blue flax fields and the yellow rape fields, hoping that we will all get to see that sight sometime, in Canada, if not on our own garden farms.

CHAPTER ELEVEN

Legumes

The Overlooked Partner in Small-Scale Grain Raising

A farm or garden, even the best ecological farm or garden, is essentially an assault on nature. You carve out a plot of ground and grow upon it what you want to grow, not what nature would have naturally provided there. Lessening the impact of that assault, by allowing the greatest number of "all creatures great and small" to live and die in mutual beneficence on that plot of ground, should be a major goal of the garden farmer.

Nature's rewards in the attainment of that end are balance and variety. The more varied the animal, mineral, and vegetative life that exists on a farm, the greater the interaction and, therefore, balance that can be achieved. For example, with a favorable climate, quail will proliferate naturally on your farm if provided brushy cover for nesting and protection. Quail can, in turn, control the chinch bugs who might otherwise ruin the outer rows of your cornfield or armyworms that might otherwise move into wheat fields from the fencelines. An unthinking farmer will cut down his brushy fencerows, which would have provided cover for quail, in order to gain four more rows of corn, and then he must spend money to spray the bugs that would ruin those four rows. Worse, the farmer must then travel to some wilder place far away to hunt quail. When he sprays he may also destroy other predators of pest bugs, thereby making even more spraying necessary.

Each life and death up and down the food chain directly or indirectly supports the others while at the same time preventing any long-term domination by any one species. Nature abhors an excess as much as she abhors a vacuum. She obeys, unerringly if blindly, the basic dictum of self-preservation: *In equilibrium lies survival.* This is the essential principle of organic farming. It should be the essential principle of foreign policy, too, instead of the suicidal notion now in vogue that bombs are the way to self-preservation.

So when you think of grains, don't isolate them as modern agribusiness does with its gigantic fields of grain. Think of grains as just another link in the food chain leading from the smallest microbe in your soil to the biggest animal in your barn and the healthiest mind and body in your family. And understand that you are not just increasing the variety and balance of your gardens or fields by growing grains, but exponentially increasing all the species of plants and animals that give sustenance to your grains as well as feed upon them.

Since grains are cereals and cereals are grasses, their introduction leads rightfully, almost inevitably, to the growing of legumes, with which grasses have a symbiotic nutritional relationship. This relationship is best illustrated by the way grasses and clovers grow together in a pasture field or lawn. Given proper moisture and soil pH, white clover will volunteer almost everywhere in America. As it grows, the clover draws nitrogen from the air into the soil beyond what it needs itself. As the soil becomes enriched with this nitrogen, bluegrass, which will also volunteer almost everywhere, is encouraged to grow. Eventually, the grass dominates the clover until it uses up the readily available nitrogen. Then the clover comes on strong again until sufficient nitrogen is available to initiate another lush crop of bluegrass. If not overgrazed, the grass and clover maintain a dense enough cover to compete well with weeds. This natural rotation—dominant grass to dominant clover and back again—prevents diseases and insects that thrive on one or the other plant from overpopulating, since they necessarily must subside with their host plant. In other words, a certain natural, dynamic equilibrium is established. Add grazing animals, in moderate numbers, and an almost permanent equilibrium can be maintained, because livestock will control most weeds and volunteer tree seedlings that would, in humid areas, finally turn the pasture into woodland. The farmer then has only to control certain weeds, thorns, and thistles that livestock won't eat.

The natural rotation between white clover and bluegrass can be used as a model for your managed rotation of grains, legumes, and vegetables. In summary, legumes in rotation can:

1. Supply 100 pounds or more of nitrogen per acre per year to your soil through nitrogen fixation.
2. Add to the soil as green manure as much as 4 or 5 tons of organic matter per acre per year, and in some cases 50 or more pounds of potash per acre per year.
3. Act as sanitary agents by interrupting disease and insect cycles of grains, fruits, and vegetables.
4. Choke out weeds that become established in previous row-crop vegetables and grains.
5. Provide an abundance of protein food for you, your animals, and your soil microbes while they are accomplishing the first four tasks.
6. Spread your available hours for work over a longer season. Legumes growing in rotation with grains, fruits, and vegetables will allow you to stretch the same number of hours over a longer portion of the year instead of having too much to do for three months of the year and not enough the other nine. In other words, you produce more while working at an easier pace.

Legumes can be divided into two classes for convenience's sake: clovers and beans. Clovers for our purposes here are mainly alfalfa, red clover, alsike clover, white clover, and, in the South, crimson clover. There are others, of course, perhaps one that would fit a specific area better than the ones I have mentioned. Even locust trees are legumes, and there has been much discussion and research into the idea of combining grass and fine-leaved thornless locust trees for a sustainable pasture. The beans of the locust also provide high protein feed for grazing livestock.

In the bean group, we're talking about all the dry beans already discussed along with garden peas and beans. All these foods can also add nitrogen and green manure when rotated with other vegetables. Think of clovers as plants for longer-term rotations and beans and peas as plants for short-term rotations.

Nitrogen Fixation

I have already addressed the subject of nitrogen fixation in previous chapters. However, with the price of manufactured nitrogen fertilizers skyrocketing, it bears repeating here. The amount of nitrogen any particular legume can fix in the soil varies, depending on many factors, and even experts do not like to use explicit figures. But in general, with all conditions right for good microbial action, plants like soybeans or cowpeas can draw 100 or more pounds of nitrogen from the air annually and make it available to a following crop. Alfalfa, which is considered by most farmers to be king of clovers, has been rated on occasion even higher; it can fix over 180 pounds of nitrogen per acre per year, though usually 100 to 150 pounds would be considered excellent. Other clovers can be counted on for at least 100 pounds when conditions for bacterial action are favorable. When planting clovers or other legumes, inoculating the seed with *Rhizobium* bacteria will often enhance nitrogen production, as pointed out earlier. And when conditions are right for vigorous clover growth (when soils are well-drained and not acid) *Azotobacter* bacteria, which fix nitrogen on their own—without legumes—will also be more active.

At any rate, the 100 to 150 pounds of nitrogen you gain saves you the cost of that much nitrogen you'd otherwise have to buy.

(Nitrogen fertilizer tripled in price in 2008, as this new edition was being prepared.) Legumes also save you the considerable labor and cost of applying fertilizer.

The Value of Green Manuring

Alfalfa in a year's time can produce 4 to 5 tons of organic matter per acre if all the hay is mulched and returned to the land rather than fed to animals. However, if it is fed, the manure returned to the soil is even more valuable. And that's not counting the value of alfalfa's roots, which can penetrate deep into the soil and bring to the surface minerals and trace elements previously leached away by erosion. Also, alfalfa plowed under returns to the soil at least another 100 pounds per acre of nitrogen and nearly that much potassium. And other legume clovers are almost as beneficial as alfalfa. Sweet clover produces even heavier tonnages of organic matter, but I don't recommend it. It takes a heavy plow and lots of fossil-fuel power to turn it under. It also can be toxic as hay if it is not cut and dried properly. On the other hand, red clover makes an excellent green-manure crop. Because the alfalfa weevil makes alfalfa problematical where I farm, I prefer red and white clovers, but farmers don't all agree with that. Farmers don't all agree on anything.

Both soybeans and cowpeas, as mentioned earlier, are good green-manure crops, though the former is so commercially valuable as a harvested bean that it is seldom used that way today. Any bean or pea will do almost as well. They will give you a green-manure crop in two months. Clovers take a year at least.

Continuous cropping of the same or similar crop is risky business because it favors a concentration of pest insects and diseases. You can avoid some disease and pest bugs just by rotating vegetables of different families among themselves or with grains. But putting legumes in the rotation gives you even longer time lapses between the various crops. This helps not only to control bugs and diseases, but also persistent weeds like chickweed.

Clovers for hay or pasture will give you weed control if managed for that. Beans and peas won't. With alfalfa, the control of weeds comes mostly from repeated mowing for hay along with the heavy growth after mowing shading out weeds. When the alfalfa crop is cut three or four times a year for hay, and that practice is followed for several years, excellent control of most weeds is obtained. Then a follow-up crop of corn, for instance, is almost free of weeds for the first month or so after planting, when weed control by cultivation is difficult for organic growers because the corn plants are so small at that time that the cultivator or hoe might cover them with dirt. The more fertile the soil, and the lusher the alfalfa, the more complete the weed control. If, however, alfalfa grows four or more years on the same field continuously, it will thin out, and weeds and grasses will gain a footing. So it is best to rotate the field to something else before returning it to alfalfa. Do not reseed alfalfa directly after alfalfa. The new seeding will not grow well. Also, if you are taking all the alfalfa off as hay, you must apply additional potassium fertilizer.

Red clover works almost as well as alfalfa as a weed suppressor, but will have strong growth only in the year after seeding. But, when properly managed as pasture, it can reseed itself year after year almost as well as white clover. If it doesn't, you can sow new seed in an old stand. I have been doing this for quite a few years now and it does work, although the stands are not as good as if an intervening crop had been planted for a year or two. My best results in this regard (I am trying to eliminate all cultivation from my rotational pasture system) is to broadcast 'Alice' white clover (an improved white clover variety that grows a little taller than wild white clover) into a declining red clover stand.

Legumes for Food

Jane Jacobs, in her interesting book *The Economy of Cities* (Vintage Books, 1969), observed that alfalfa was grown in French city gardens in the Middle Ages—a century before it was widely adopted in rural agriculture. It is believed that the Moors, even earlier than the Middle Ages, recognized the ability of alfalfa to make land more fertile. But alfalfa was no doubt grown primarily in gardens for its herbal value, and again herbal folklore seems to have some basis in fact. Alfalfa is rich in nitrogenous compounds that are thought by some nutrition scientists to inhibit cancer. Red clover tea—made from blossoms dried in summer—is considered a blood "purifier" and has been used as a folk remedy for cancer.

Be that as it may, alfalfa and clovers are rich in protein. Good legume hay can supply all the protein a cow needs. Many people seeking to vary the kinds of protein in their diet take alfalfa tablets. Futurists even believe that alfalfa could be one of the more practical solutions should a worldwide protein shortage occur.

Examples of Rotations

The best way to appreciate the economies of labor gained by rotations with legumes is to consider some actual examples. My own original plan was set up to supply all our own vegetables (except those grown in our little green-garden near the house), plus grain and hay for at least one cow, one hog, thirty chickens, and fifteen rabbits per year, and some for the family too, while at the same time maintaining a soil-fertility level that required only a minimum amount of additional purchased fertilizers.

With this rotation plan fully operative, the area that was maintained in the rotation measures approximately two acres, split into six quarter-acre plots. Table 6 shows the plan, beginning in the fall and carrying it through five seasons.

The rotation moves from A to F and back to A again, as you can see. In other words, what happens one year on plot A will happen the next year on plot B. What I have is a six-year rotation—an unusually long one even in traditional agriculture. Agronomists would formalize the rotation as corn, corn, bean-vegetables, wheat, hay, hay, or C-C-B-W-H-H.

Obviously, I did not start with a full-blown six-year rotation. One must start somewhere, and the usual situation on a garden farm is to start with a field that has been in sod. In my case, the first year I plowed all the plots except F and planted them to corn interspersed with a few rows of garden produce. The first fall I harvested the corn early from plot A and planted it to wheat, interplanting in spring to alfalfa. Plots C, D, and E went that spring to corn, again interspersed with garden produce. F remained in sod. I planted B to soybeans in June, plowed under in August, and planted wheat in the fall. This plot was seeded to alfalfa the next

TABLE 6. SIX-PLOT ROTATION PLAN

A (¼ acre)	B (¼ acre)	C (¼ acre)	D (¼ acre)	E (¼ acre)	F (¼ acre)
FALL					
Second-year alfalfa	First-year alfalfa	Plant winter wheat; apply lime	Fall-plow for next year's vegetables and dry beans	Fall-plow for next year's corn	Sod
SPRING					
Second-year red clover or alfalfa	Apply potash on first-year red clover or alfalfa in April	Broadcast clover early in March; apply phosphorus in April if necessary	Plant garden mulch tomatoes, potatoes, melons in June	Manure, if available, or purchased fertilizer; plant corn in early May	Plow down green manure; plant corn in late May
SUMMER					
Make hay	Make hay	Harvest wheat; gather some straw; clip clover seedlings	Harvest garden produce in August; incorporate spent plants as green manure	Weed-cultivate corn	Weed-cultivate corn
FALL					
Make hay	Let late red clover or alfalfa grow; do not make hay	Let red clover or alfalfa seedlings grow	Plow under remaining vegetable residue and mulch and plant wheat; apply lime	Harvest corn early, removing both fodder and corn for feed	Harvest corn late, leaving stalks on ground for organic matter
SPRING					
Plow under alfalfa for green manure plant corn in late May	Second-year alfalfa	First-year red clover or alfalfa; apply potash in April	Broadcast clover seed early in March; apply phosphorus as wheat greens up in April	Plant beans and vegetables; manure or mulch tomatoes, potatoes, etc., in June	Plant corn in early May; apply nitrogen before or after

spring, as plot A was the previous spring. By fall of 1976, with the wheat crop removed from plot B and the alfalfa seedlings making a good stand, the scene was set for the beginning of the rotation schedule as shown in table 6.

This six-year rotation could as easily be a five-year rotation. I added another year of corn (plot F) because at that time I foresaw needing more corn. I could get the same amount of corn on a five-year rotation by reducing the amount of hay in the rotation. If instead of alfalfa I grew red clover, as I did later, I would necessarily reduce the amount of hay in the rotation, because red clover will not produce a good yield of hay the second year after planting and should be plowed up. With red clover, the rotation would be corn, corn, beans, wheat, clover, and then back to corn.

I could also achieve the same total production of food by making the plots larger but fewer in number, thus corn, beans-vegetables, wheat, hay, and back to corn. I prefer six smaller plots to four larger ones to spread the workload more and to allow myself the prerogative of working other crops conveniently into the rotation.

This rotation maintains fertility almost automatically. In any given year, two plots will be in alfalfa (or, what I've learned now, after thirty years, red clover), one plot in beans plus heavily composted vegetables, and the other three in grains. Three plots are adding more fertility than they are taking out, and three are taking out more than they are adding. A sort of natural equilibrium is established, and I can build overall fertility by adding manure and small, economical doses of additional fertilizer that I can afford. Organic fertilizers with a guaranteed analysis of nitrogen, phosphorus, and potassium are available and ideal for my situation, but too expensive for me to use in large amounts. My "bias" is that an organic farm must be producing its own animal manure for fertilizer if it is to be profitable.

To understand how conveniently a rotation spreads your workload, I will describe in some detail how I managed mine. (I'm not all as efficient as this may sound, however. Sometimes I simply stand and stare into space instead of working.) Any year will suffice for an example, but let's go through the year that I sketched out above.

My work year begins in plot C in March, as yours will too if you live in the same latitude as I do. Farther south, you start earlier, of course. I try to pick a quiet morning when the ground has frozen just slightly. With a plot size of a quarter acre, I need about 4 pounds of alfalfa or red clover seed in my seeder hanging from my shoulder. The wheat is still dormant on the plot, and the clover seed flying from the spinner of the seeder easily finds its way down to bare soil, falling most often into the tiny surface fissures created by the frozen soil crust. Later in the day, or tomorrow, or very soon at any rate, the soil will thaw, flowing together and giving the tiny seeds all the cover they need. Then, when the soil warms up, they will sprout and grow. My first planting of the season, and the easiest, is completed in about an hour. It would take less time, but I stop often in March to listen for spring birdsong or (today, when I am older) to catch my breath.

In early April, or whenever the ground has thawed and firmed enough so I can walk on it without slopping mud up to my elbows, I apply potash to plot B and phosphorus to plot C. I just walk along with a bucketful and hurl it on the ground by the handful, as evenly as I can. By now the wheat is just beginning to green up and I can hear the birds singing without having to stop and listen.

Next I move to plot D as soon as the soil is dry enough, work it with my tractor and disk harrow, which came out of the factory shortly after Methuselah died, and plant peas and potatoes. Then it's on to plot E in early May to work the soil and plant early sweet corn and field corn. If I have manure or other fertilizer I'm not going to need for vegetables in plot D, I'll apply it here before planting the corn.

Next I plow under the green manure on plot F and plant more field corn if I didn't get enough planted earlier, and late sweet corn. If the weather has been normal, this should be about May 25. Then I hurry to plot D, where I plant garden beans and tomato plants. The rest of plot D I keep disking to hold down the weeds. Later I will plant dry beans here.

By now it is June, which I spend in plots D, E, and F, hoeing weeds and mulching. In late June, I move to plots A and B and take off the first cutting of hay. I try to mow the alfalfa just as it begins to bloom, when its nutritional value is highest. When the hay is dry I rake and haul it to the barn or stack it in the field. It makes excellent mulch too, if I have more than I need for feed.

In early July, the wheat is ripe in plot C. I harvest and gather some of the straw for bedding or mulch. I clip the stubble with a rotary mower and let that straw lay right on the ground for fertility, organic matter, and a mulch cover for the alfalfa seedlings. It doesn't hurt to clip those seedlings. Seems to make them grow back healthier and stronger.

Next I plant the soybeans in plot D, and continue weeding whenever I can't think of an excuse not to.

In late July, the alfalfa should be cut for hay again. If I wouldn't need it all for hay or mulch (I always do) I would mow it with a rotary mower, which will shred the plants back onto the ground to provide excellent mulch and fertilizer value. This would be rather wasteful of a good protein feed, I suppose, but, in an organic sense, nothing returned to the soil is wasteful.

August is harvest month in the garden (plot D). I will also want to go fishing and sit in the shade a lot.

In September, I plow under or rotary-till the soybeans and other spent beans in plot D that we didn't harvest for food. After frost, I do the rest of the plot the same, and toward the end of the month I plant the whole thing to wheat. Meanwhile in plot E, I cut the corn, stalks and all, tie it into bundles, and shock the bundles together. I want not only the corn, but also the fodder for rabbit and livestock feed, and for bedding.

From the corn in plot F, I harvest only the grain. The fodder is returned to the soil to decompose and build organic matter. In October, I shuck the ears by hand and toss them in the truck, moving down the row of stalks methodically unless I spy a flint arrowhead on the soil surface, which happens regularly in this area. Then I stand there awhile contemplating the artifact and wonder if its owner planted corn here too. Afterwards, I shred the stalks with a rotary mower, and plow or rotary-till them under in the fall to control corn borers.

Meanwhile, before hard frost, I *could* take another cutting of alfalfa from plots A and B. I probably will, from plot A at least, since I will be plowing it under the following spring rather than keeping it for another year of hay. Plot B, on the other hand, will be in hay next year, so I want it to winter over in good shape. Not cutting the last crop means leaving a good cover of dead alfalfa on the soil surface over winter, which helps protect the roots from heaving (and gives quail and other wild things a place to hide). It also means that the alfalfa next spring will get a strong start.

In November, I spread lime over the new wheat on plot D.

With this rotation, I can produce quite a bit of food with my spare-time labor and mostly primitive hand methods. Long-term fertility management sort of takes care of itself. Every year one plot receives extra potash; one plot rock phosphate; another lime; two get green manure; and at least one gets animal manure and compost.

There can be all kinds of variations on this rotation plan. One of the corn plots could just as easily be oats. Barley could substitute for wheat. Buckwheat could be grown after either oats or early peas. Strawberries could be worked into the rotation on a two-year basis, as part of one of the corn or alfalfa plots. By using temporary electric fencing, I could put a pasture into the rotation too, planting a timothy–red clover pasture or a ryegrass–ladino clover pasture instead of alfalfa and letting livestock graze it instead of making hay. Or make hay off one plot and graze another.

High-value cash crops can be worked into a garden-farm rotation where the plots are small in size like mine. Where tobacco is grown, a quarter-acre allotment could be part of such a rotation. Or a quarter-acre of tomatoes, potatoes, strawberries, melons, or even flowers to sell. When the price of good hay soars out of sight, as it regularly does these days, even your alfalfa begins to take on the glow of a high-value crop.

Needless to say, by 2008, when I revised this book, things had changed somewhat to accommodate old age and hopefully a little more know-how, although not the principal rules behind the rotations. Now I mostly rotate livestock (sheep) from one pasture plot to another, letting them do the work of fertilizing and harvesting, and rotate only corn and red clover as described in chapter 2. The only plowing is for the corn, and if I had a heavy rotary tiller for the tractor, I'd use that instead. The gardens proper we have divided into four plots that rotate yearly: (1) sweet corn; (2) lettuce-beans-beets-peas, carrots; (3) potatoes-sweet potatoes, onions, squash; (4) tomatoes, peppers, eggplants, melons; with more or less permanent plots of asparagus and strawberries and raspberries. The berries are rotated from one side of their plots to the other yearly.

Often Overlooked Details about Clovers

Clovers can either be sown by broadcasting on top of the soil during the winter, as I do, or in spring, drilled about a ¼ inch deep in tilled soil, or with a no-till planter in sod. In some parts of the country clovers are sown in August on tilled soil if moisture is plentiful. Where winters are severe, August plantings can be risky, and plantings later than that even riskier. Seeds that sprout in September are highly likely to winter-kill.

Clover seed can be stored for years and still give good germination when planted. At farm sales, clover seed that has been hanging in a dry barn for several years will still bring a good price because farmers know it will grow. Not necessarily so with grass seed.

When red clover seed is broadcast in dormant barley or wheat or drilled in with oats or by itself in spring-planting season, it grows perhaps only 5 inches by the middle of the summer. You can barely notice it in a dry summer if the wheat or oats crop in which it is growing is heavy. But after the grain is harvested and the stubble has been clipped or grazed for a short time, the clover spurts up to about a foot tall by winter if rain is ample. That can make good winter grazing. The plants look dead by December, but they are still nutritious, and if they are standing upright (not yet flattened by winter snow) they provide good grazing. Needless to say, in a drought year, spring-planted clover seedlings are liable to die. I have had that happen twice in thirty-five years.

When clover seed is threshed with a combine, as described earlier, it still contains much chaff and often weed seeds. If not run through a seed cleaner and then dried, the seed will heat and deteriorate to the point where much of it won't germinate. So, after cleaning the seed, spread it out indoors—no more than an inch thick—and allow it to dry for a month. Then it should be safe to bag.

Red clover seed can be harvested by combine throughout most of the United States. Alfalfa needs long periods of dry weather to produce a harvestable seed crop, and most of it is grown for seed in the western states. I have never seen it written down anywhere, but when I was a boy and red clover was regularly harvested for seed and sold, the bane of the clover fields was narrow-leaved plantain—what we called buckhorn (also called ribgrass). It is now the bane of suburban lawns. The seed was very difficult to separate from the clover seed in seed cleaners of that time, and so any batch of clover seed infested with it would not bring a good price. I remember one very agonizing day when our family crawled over our fifteen-acre red-clover field on our hands and knees with pocketknives to cut out the buckhorn. Dad only made us do that once, thank heavens. He wouldn't have done it once except that Grandfather, who owned the farm at the time, insisted. Now all these many years later I know how witless that weeding was. Narrow-leaved plantain used to be planted deliberately by English farmers in their pasture mixes because it is nutritious, has medicinal value, and sheep love it. It is also practically immune to dry late-summer weather when pasture is at a premium. Today my red clover, to both my own and my sheep's delight, is full of plantain, both the narrow-leaved and broad-leaved versions.

Finally, the rewards of combining legumes and grains on your garden farm can extend directly to your table. Alfalfa and clover sprouts both make great salads.

I can't resist pointing out that the old herbals, going back to China in the sixth century B.C., insist that plantain leaves make a marvelous poultice for wounds, stings, and burns. And, as I wrote earlier, red clover seed is revered in tradition as a cancer fighter. When you work with nature rather than against her, all things (well, almost all things) are possible.

CHAPTER TWELVE

Feeding Grain to Animals

At the risk of repeating a few things mentioned earlier, I want to discuss feeding grain to animals in more detail. In a way, I hate to broach the subject, because in doing so there is no way to avoid stirring up the commercial animal-feed businesses over what I consider their myths about animal nutrition. If you think there is a decided difference of opinion among experts over human nutrition, just multiply that difference to the fourth or fifth power in the field of animal nutrition. Humans can say anything about what animals think about their nutrition because the animals can't talk back to them. We are dealing in this regard with a religion, the Worshipful Brethren of Corn and Soybeans and, heaven knows, religions don't like to be contradicted.

The desire to sell something colors the whole animal feeding business—well, it colors the whole notion of business. Not only are there many commercial companies involved, all with certified nutritionists on their staffs proclaiming their feed to be the best, but there are hundreds of farmers like me, all self-proclaimed experts with theories of their own on the best feed formulas. Since all farm animal feeds use the same basic ingredients, all manufacturers speak with a measure of truth. Just remember that animals lived unnumbered centuries without benefit of any of us.

Another way to look at feeding livestock on your garden farm is to compare the process with feeding babies, crude as that might sound. Some folks believe it is more convenient to buy a variety of canned baby foods at the supermarket, and believe, at the same time, that they are reasonably assured that baby is getting nutritious fruits and vegetables. But that does not mean, as commercial baby-food manufacturers would like for us to believe, that mother can't prepare her own baby food as nutritionally good or better than what she can buy.

If grain mills for animal feeds were as inexpensive and easy to operate as baby-food grinders, there would be little reason at all for garden farmers to buy commercial feeds, except to save time and labor. Despite innuendos from feed salespeople to the contrary, farmers used to (and some still do) grind their own grains, add their own supplements and minerals, even vitamins if they think necessary, with quite satisfactory results. And some don't believe that milled grain is any better than unmilled grain. Gracious me, what heresy!

Yet the first time you go to a feed store to try to buy feed or to get some advice on feeding your animal, most often the person in charge will seem to believe (I'm convinced some really do believe it) that an almost mystical health value attaches itself to a commercial sack of feed simply because it comes from a commercial feed company, or because it has passed through a commercial grinder. The contrary garden farmer feeding homegrown whole grains and homegrown processed feeds is considered a witless apostate and his animals will all wither away. It is as if the animal nutritionists who work for the commercial feed mills hold secrets of healthful food that the rest of us are not privy to.

Within the framework of modern confinement feeding of farm animals, that mystic faith has some justification. If hogs live their entire lives on concrete and are fed through augers one diet of milled hog feed their entire lives, then that feed better contain every known mineral, vitamin, protein, and carbohydrate that the hog needs. And the fact that confinement-fed animals still *do* suffer disorders and disease proves that the scientists haven't yet solved all the mysteries involved.

The Garden Farm Difference

The garden farmer's hog lives an entirely different kind of life. It may have the freedom to roam a field or large lot where, by rooting in the dirt and eating a variety of natural foods, it balances its own nutritional needs quite well. If the homestead hog is confined to a smaller lot, as mine is, it still receives a wide variety of feed, including our table scraps, which contain nearly all the nutrition my family seems to need, so why not Mr. Hog? He also gets a fistful of fresh clover or clover hay, the most complete natural food the farm produces, according to Ohio State University agronomists. In addition to corn and wheat and all kinds of garden residues, the hog gets an occasional acorn or hickory nut, an apple or a melon rind, and he can always root in his lot for grubs. Does this lucky animal need the magic potions from Purina? I don't think so.

Grains are ground for animals primarily to put the feed in a form the animals can chew and digest fast; the same reason you grind up baby foods for toothless infants. Only the animals aren't toothless by far. If the animal consumes more in a given period of time than he would roaming free-range, he will likely gain weight faster, and that's the name of the commercial game. It is also why we have such a dreadful number of obese people in our society. But that kind of feeding need not apply to the garden-farm situation except where the animal *can't* eat the whole grain. Small chicks have to have their corn at least cracked before they can swallow it; the whole kernel is too large. Young pigs may get sore mouths biting hard corn off the cob and chewing it up. Lambs and colts—and even adults of some species—may refuse to eat oats because of the tastelessness of the oat hull, until the grain is rolled or milled.

On the other hand, a hen will just as readily consume whole grains as milled ones, and the natural grit in her crop, which she gets from pecking and scratching in the sand and soil, will do the grinding and digesting. The only difference is she will not eat as much (her crop will hold more milled feed than it will hold whole grains, and therefore more total food nutrients). The garden farmer can feed mostly whole grains with satisfactory results. In fact, in many cases the homestead animal is overfed. Our twelve laying hens receive no milled feed at all anymore. They lay just about as many eggs on whole grains—or on no grain at all when they have ample room to roam through field and woodlot.

I'm more inclined to use milled commercial feeds for baby animals. At that age, intake of vitamin A and iron is critical, and not necessarily available in sufficient quantity because of the small amount of rough homestead foods they eat at that time. In nature, running with their mother, chicks get all kinds of tiny specks of food that mother hen searches out for them and puts in front of them with her beak. Motherless chicks don't have that advantage. When I was a boy, we put a trough full of loam from the woods (soil that had not been farmed and therefore not depleted of natural fertility and trace elements) in the pen with baby pigs. They rooted in that dirt and got their iron from it. If they had been on pasture, they could have gotten their own iron. Commercial growers today give the pigs iron shots as a matter of course to prevent anemia. Some veterinarians tell me that on farms operated according to an organic, balanced-fertility program espoused by "natural" farmers, iron shots can eventually be discontinued because the pigs get enough through sows' milk and foraging on their own from healthy soils. Likewise, I no longer feed lambs any commercial creep feed. They are born in spring when new grass is growing, and they learn to graze from their mothers by the third or fourth day of their lives.

Animals need carbohydrates, proteins, fiber, vitamins, minerals: everything humans need in our own diet. They will eat all the grains mentioned in this book, all legumes and grasses and most weeds, especially young, succulent weeds. Some wild plants contain medicinal ingredients the animals instinctively know more about than we do. Animals will generally be healthier grazing a slightly weedy pasture than in one where only one type of grass or two are allowed to grow. *Variety* is the key to feeding animals in the natural environment of the garden farm. If your livestock have access to many kinds of food, they will balance their diet on their own.

The only exception is if you live where soils are naturally deficient in certain essential trace elements, like zinc or selenium. Where organic matter is high in the soil, trace element deficiency is extremely rare, but under intensive farming, zinc, boron, selenium, and other trace elements may be lacking. It has become necessary in some parts of the Corn Belt to add selenium to feeds where once there was no such necessity. Some soils of the northern plains naturally contain too much selenium, too. But the likelihood of such deficiencies or surpluses being critical in your garden-farm situation is unlikely. Check with a veteran farmer in your area if you are in doubt.

Though grains will be the basic supplemental feed on a garden farm, pasture should be the major source of their food. Clovers are the "best" all-around feed not only for grazing animals like cows, sheep, goats, and horses, but also for hogs and chickens. (Though one should not use the term "best" for any one food—"best" is when a great variety of feeds are provided.) Alfalfa experts point out that clovers come close to being a complete animal feed. They contain high amounts of vitamins and protein and carbohydrates and even minerals.

Cows and sheep, don't forget, will sometimes bloat if they have free access to unlimited quantities of fresh, lush alfalfa or other clovers, or green corn, or even kernels of mature corn. A hungry or half-starved cow, sheep, or horse, unlike a hog or chicken, will keep on stuffing itself on food it really likes. So cows bloat on fresh clover and horses founder on too much grain. It is not good to give hogs a lot of whole wheat at one time, either. As I said earlier, they might swallow it whole, and it will swell up in their stomachs and give them a very bad stomach ache, or even kill them, although that is very rare. Clover pastures should also contain grass to help control bloat. Turn the animals on it gradually, after filling them with hay first. And never turn them in hungry on a pure, lush stand of green clover still heavy with dew or one that has recently been frosted in late fall because of potential bloating. Birdsfoot trefoil is an exception; it is the one clover that animals don't seem to bloat on, probably because it is very fine-leaved and fine-stemmed.

Corn is the principal grain fed to animals, though barley, grain sorghum, and even wheat can be substituted for it. Corn provides more energy per pound than the other grains, but it is low in protein. In ground feeds, a little oats may be added to the corn, along with alfalfa meal, soybean meal, meat scrap meal, linseed meal, or cottonseed meal for protein, and bone meal (or other phosphate compound) for calcium and phosphorus. Trace mineral salt is added to commercial feeds. Animals not on commercial feeds need a block of salt to lick on. The brown mineral blocks are okay for cows, but not sheep, so the feed supply dealer I buy from says. He seems most knowledgeable, so I follow his advice, and my sheep do just fine without those mineral blocks.

Soybean meal is the principal protein supplement in commercial husbandry today, though it doesn't contain the range of proteins some of the other meals enjoy. All can be purchased separately at feed stores if you wish to make your own feed mixture. Or you can tell the local feed-mill operator what mixture you want and he will do the mixing. Commercial feed salespersons usually advise soybean meal for protein supplement because it speeds up the fattening process. But you don't *have* to feed out a hog in four-and-a-half months on a garden farm. If your hogs don't reach the butchering weight you want until they are six months old, what's the difference to you? A longer period of feeding out the animal won't mean losing money, as it might for a commercial hog producer. The meat might be an itsy-bitsy degree more tender if fattened faster, but will have itsy-bitsy less taste to it, too. A longer feeding period could mean saving money for the garden farmer, in fact, by not using as much expensive protein supplement. Connoisseurs of pork, especially of smoked hams, maintain that hogs fattened more slowly (especially on a diet supplemented with acorns) produce better-tasting meat anyway.

The next most important thing to remember (if not the first thing) when feeding grains is that nutritional value varies considerably, depending on the soil in which the grain was grown, the weather, the variety, and the way it was handled and processed. This fact is what makes advising people on animal nutrition so difficult. Corn, for example, is not a standard, packaged item off an assembly line. Some corn contains more protein than other corn, *even within the same variety*. In some tests, normal hybrid corns sometimes contain more protein than special high-protein varieties, the difference being the soil, culture, and weather conditions. In processing, almost everyone today will admit that corn dried with artificial heat sometimes gets too hot and is therefore less nutritious than slowly air-dried corn. And some feeders believe that old, open-pollinated corn varieties are, pound for pound, more nutritious than the highly specialized hybrids of today. All of which means that the identical feed formula on two different farms might have a different nutritional value.

Feeding Chickens

If you really insist on buying a commercial mash for your hens, a good mix can be made from 1,250 pounds of ground corn, 200 pounds of wheat, 100 pounds of alfalfa meal; 240 pounds of soybean meal; 100 pounds of meat scraps; 50 pounds of bone meal, 40 pounds of ground oyster shell, and 10 pounds of salt, plus vitamins, if you think the latter are necessary. In my opinion, this is a better feed than if all the protein supplements were provided by one source, say soybean meal. With the variety you get a broader range of proteins, which not only means a healthier animal, but also more protein-rich eggs and meat, and a manure capable of producing plants with a broader range of proteins in them. But you can make an adequate feed with just one of the protein sources mentioned, or two. If two, choose when possible one plant source (for example, soybean meal) and one animal source (for example, meat scraps). But remember that a few hens with the run of woods and pasture can supply everything they need except in winter from foraging.

For broilers a similar mix is fine. Usually in fattening poultry of any kind an all-mash diet will do a quicker job, as already noted. Not necessarily better, but faster. I used to feed my layers and broilers the same feed: whole corn, whole wheat, whole sweet sorghum grain, whole grain sorghum grains, some broomcorn seeds (hens don't much like them), millet, and buckwheat; alfalfa hay; some grass seeds that they pecked off the stalks when on free range; weeds gone to seed; sour grass (high in vitamin C) when the hens were penned up; plus some eggshells, oyster shells, a little bone meal sometimes, table scraps, garden wastes of all kinds, and a little commercial mash in winter.

We feed new chicks a little bread and milk if they are acting unsatisfied, let them run on the lawn where they chase bugs and nibble at clover after they are about a week old, and keep commercial feed beside them at all times, for reasons mentioned earlier. We don't get chicks from the hatchery until June, so we don't need to have a brooder house to keep them warm. We just keep them in a big cardboard box with a light bulb suspended over them for about two weeks. Then it's off to the coop with them. Standard advice is not to bed chicks with sawdust. The theory is they may eat some of the wood particles and die from ingesting too many of them. However, we have raised chicks on sawdust, wood shavings, straw, peat moss, and newspapers, and I can't say that the mortality rate is any different with any of these beddings. It seems that a chick or two in every batch always dies. The most important detail to survival, we think, is to keep the temperature in their "brooder box" at not less than 90°F for the first few days.

Feeding Rabbits

Rabbits can be raised on very little grain, but, because rabbits are usually raised in confinement, feeding grain is usually more convenient. (My grandfather raised rabbits in a fenced lot. I am sure that would work because I once lived in the countryside where a neighbor had let his domestic rabbits escape their pens. These rabbits proliferated in the wild until they were hunted down. It was startling to scare one of them out of a brush pile.) Commercial rations in pelleted form contain alfalfa and other roughages, oats, wheat middlings (about the same as wheat bran only further refined to contain less fiber and more protein and fat), corn, or barley. I feed very little commercial feed, and only to weaned young animals that are going to be butchered. They live mainly on top-quality alfalfa hay, which they love, and a little whole wheat and corn. The corn is a bit too hard for them to manage well, and they drop some of it through the cage. Hard flint corns should not be fed to rabbits at all unless milled. The kernels are too hard. I have fed oats, soybeans, millet, and sorghum seeds to rabbits with good luck, giving it to them stem and all. The rabbits nibble the grains out of the hulls themselves and like to chew on the straw. With tough fibrous material like that to chew on, rabbits do much less gnawing on their nest boxes.

Feeding Cows

A dairy cow's grain ration can be very simple if her hay has a lot of high-quality alfalfa or red or white clover in it. In fact, in some tests, cows fed a very high-quality legume hay did fine without any grain at all. But if your hay is not that good (or if you have been brainwashed by the Worshipful Brethren of Corn and Soybeans), to every 1,000 pounds of ground corn and cob, add 10 pounds of bone meal and 10 pounds of salt. Instead of all corn, you can mix 700 pounds of it with 280 pounds of ground or rolled oats and add

200 pounds of soybean meal to get the necessary protein in the mix. A more nutritional concentrate for calves, if not cows, would be a formula of 21 percent corn and cob, 20 percent oats, 15 percent wheat bran, 11 percent soybean meal, 10 percent linseed meal, 5 percent alfalfa meal, 10 percent dried whey, 5 percent molasses (calves like candy just like kids do), 2 percent bone meal, and 1 percent salt. To that mix a commercial dairyman might add 2,500 I.U. of vitamin A and 300 I.U. of vitamin D. If you want calves as healthy as your children, and you think your children need vitamins, then you will probably want to give vitamins to your animals too. You can tell by the way I use "if" and "probably" that I don't think people eating good, naturally raised fresh food need dietary supplements. But when I say that around some nutritionists they get very upset.

After a calf is six months old, it should be able to thrive on a ration of good pasture or hay with very little grain. If the pasture or hay is not so good, give 3 pounds of grain daily. Two months before the heifer calves, start increasing the grain ration up to 8 pounds per day at calving if pasture is not first-rate.

Milking cows should be fed grain (along with roughage) according to how much milk they give. A cow producing 30 pounds of milk a day should get about 10 pounds of grain or high-quality hay equivalent; if 40 pounds of milk, 15 pounds of grain; 60 pounds of milk, 25 pounds of grain; 80 pounds of milk, 35 pounds of grain. This is *not* a hard and fast rule.

A beef steer or heifer for fattening can run on good pasture for eight to ten months with only a little grain fed on the side, increasing the grain in the last month before butchering to give the meat a taste that the corn-fed American palate prefers. On the other hand, our last beef steer got no grain at all, just high-quality pasture and hay, and we found the meat tasty and reasonably tender. But animals fed entirely on grass and clover will rarely meet the Choice standard, let alone Prime, of our nation's weird grading system. Grass feeding makes the fat yellow, say the graders. If anything, yellow fat means the meat is higher in beta-carotene (a.k.a. vitamin A, from grasses and other plants) and is therefore more nutritious.

Years ago, a good neighbor farmer told me he had finished out a carload of beef on hay and corn in shocks—feeding the shocks, ears, stalks, husks, and all—and he'd received an excellent market price for the animals. Again, the lesson is that there are no hard and fast rules.

Feeding Hogs

A standard formula for a ton of hog feed is 1,500 pounds of ground shelled corn (you don't grind the cob along with the kernels for hog feed, as you do for cow feed) and 500 pounds of protein supplement, in the form of mostly soybean meal plus a few meat scraps and some alfalfa meal. Another mixture, especially for younger pigs, would be 1,200 pounds of corn, 50 pounds of oats, and 400 pounds of soybean meal; or 1,700 pounds of corn and 300 pounds of barley—plus trace minerals added to either mix if necessary. A little bone meal or other source of calcium and phosphorus is good for pigs too.

To feed a hog to 200 to 220 pounds butchering weight takes about 12 bushels of corn or its equivalent in other grains plus the supplement. I feed my fattening hog about one-third ground feed like that described above, and the other two-thirds whole corn and other homegrown feeds including good hay, as previously described. When we were milking a cow and had more milk than we could drink, we fed it to the pigs. That's the way to make really delicious pork.

Feeding Sheep

On good pasture, sheep require very little if any grain at all. I used to feed ewes a little oats daily for a week before breeding them. During the last third of her pregnancy a ewe can be given an average of about ½ pound of corn or oats per day, starting at about ¼ pound and working up to a pound by lambing time. But now that we have good pastures, I don't feed any grain at all—not to ewes or even to fattening lambs. Anyone who tells you differently just doesn't know what good hay is. They are feeding hay that looks like bales of used drinking straws, so of course they have to feed grain. Lambs born in April and May (as I would advise all garden farmers to arrange) do not need creep feed. (Creep feed is any finely milled grain that new lambs or calves are allowed free access to while they are nursing. Obviously it has to be fed in an area where the mothers can't get to it.) The lambs will get all the feed they need from good pasture and milk. As I have said, they learn to eat grass on pasture from their mothers in about three days.

Feeding Goats

Goats are browsers rather than grazers. That is, they prefer to nibble on tree and weed leaves at about head height, not with their noses down close to the ground. However, they will readily adapt to grazing low herbage if nothing higher up is available. Goats can be fed similarly to sheep, except for lactating does, whose needs more closely resemble those of a cow. The amount of homegrown grains you can use versus commercial feeds follows the same guidelines as I have given for other animals, with a decided bias in favor of the homegrown stuff. Just think about the millions of goats in third world countries who sustain the human population without one ounce of commercial feed. In this regard, the *Gopher Goat Gossip*, a newsletter of the Minnesota Dairy Goat Association, supports my experience. The newsletter reported in the 1970s a feed formula that was used successfully on a Minnesota goat farm, as follows: 16 pounds of whole shelled corn, 16 pounds of whole oats, 8 pounds of whole wheat, 4 pounds of soybean meal, and 3 pounds of molasses. Feed ½ pound daily to adult goats plus ½ pound for every pound of milk the goat is producing. Feed kids ¼ to ½ pound daily. In summer, with decent pasture, cut the ration by half. Needless to say, I think that when you have high-quality hay and weeds and brush for the goats to nibble on, you don't need anything else.

Speaking of good hay, it is conceivably possible (though not very probable) that you could feed *non-lactating* goats (or other farm animals) too much high-quality alfalfa. Alfalfa—good alfalfa—has a high calcium content. Male animals and dry females may not be able to handle all that calcium if fed a diet of too much alfalfa for a long period of time. Use common sense, and don't overfeed. When I have "super hay" as I call it, I feed the animals a little straw or hay that was rained on before getting dry and in the barn. If they will eat it while also having access to good hay, you know they need it.

Feeding Horses

The usual danger, especially in the case of teenagers and their pet horses, is feeding too much grain rather than too little. This can also happen and usually does, with pet cows, sheep, rabbits, and chickens. A saddle horse will do fine on good pasture with a little oats on the side. In winter, feed daily 6 pounds of oats, 1 pound of corn, ½ pound of linseed meal, and hay. Or, feed good hay. Rolled oats, rolled barley, and wheat bran are other good horse supplements.

The experts say that a mare with a colt needs 6 to 8 pounds of grain daily—eight parts oats, one part corn, one part wheat bran—and mixed grass-clover hay. The only reason I might agree is because so many horse owners seem to have no idea at all of what good pasture and hay look like. They keep their poor horses penned on a lot so small that it will hardly provide enough pasture for a rabbit. Soon the poor animal is chewing on dirt, desperately trying to gnaw the grass roots out of the ground. Horses in this situation will indeed need a heavy grain diet, but not to their advantage. Given too much grain, they will founder. Conversely, a horse that's been half-starved all winter and then turned out on lush green pasture is liable to founder too. Give a horse three acres to graze, divide that pasture into three sections so you can move it on fresh grass every ten days or so, and introduce it slowly to lush pasture in the spring. You will make up the price of that land by not having to feed as much expensive grain and hay and by not paying as much in vet bills.

I quote from the sixth edition of the *Midwest Farm Handbook* (published by Iowa State University in 1964): "When horses are fed a variety of feeds, including a legume hay, the mineral content of the ration is usually adequate." That is usually true in garden-farm situations where the horse owner provides adequate pasture, not just an exercise lot. And it should relieve you of the fear that you aren't feeding your horses correctly just because you are not spending a lot of money on special commercial feeds.

Feeding Turkeys

I think I'll continue to advise garden farmers against raising turkeys, even though some reviewers of a previous book of mine considered me negative in attitude for doing so. Turkeys are touchy to raise indoors and risky to raise outdoors. But, if you want to try it, you'll get a lot of laughs out of this clown of the poultry family, assuming it stays healthy. My brother raised a turkey. Once. One morning he couldn't find the silly thing. Finally he spied it up on the roof of his house, possibly surveying the horizon for signs of the coming apocalypse. Turkeys are by nature very apocalyptic. If they get their feet wet, they think they are drowning and might keel over and die of pneumonia. How can you blame the thing since its whole purpose in life seems to be to grace human tables at Thanksgiving? A good epitaph to put on a turkey's grave is: "This bird expected little from life and was rarely disappointed."

Feed turkeys the way you would chickens—obviously, give a turkey more grain if it's destined for Thanksgiving dinner than if you're keeping it for breeding. Turkeys are tremendous foragers if they are allowed to run loose, and that is part of the problem. They voraciously eat insects that harm your garden, but they like vegetables equally as well, especially corn. Since wild turkeys are now increasing in population and becoming very destructive of crops (again, particularly corn), I urge all garden farmers to forget about raising these flying manure spreaders and get their Thanksgiving bird from the wild. Any enemy of sweet corn is my enemy too.

Geese and Ducks

As long as they have access to green plants and some water, geese and ducks will get through a summer without any grain at all, but a little extra shelled corn won't hurt them, either. When fattening the birds for market or fortifying them up for a season of egg-laying, feed a little grain if you want to be sure of more eggs. Goslings and ducklings can be given chick feed to start them off in spring if they don't have mothers to teach them how to forage. Then, after a summer of free range, they'll need shelled corn in late fall and into winter. In January or February, in anticipation of the coming egg-laying season (or to fatten them up for eating if you did not do so in the fall) start them on a mixture of ground corn, ground oats, and ground alfalfa hay with soybean meal added to make a 16 to 20 percent protein ration. If they have a pond to dive in, they can get by on much less than that, even to some extent in winter. Look at those wild Canada geese if you don't believe me. They do just fine turning golf courses and farm-pond shores into hog wallows, maintaining a size only a little less than that of a Cessna without any extra protein at all. They do it by pigging out on the vast acreages of corn, wheat, and soybeans from sea to shining sea.

Feeding Cats and Dogs

Both these animals are by nature carnivorous, and don't ordinarily eat plain grains. I have noticed our cats occasionally eat a bit of milled corn. I think they do so because they see the chickens eating it and believe they are missing out on something. I suppose that cats and dogs could derive nearly all their nutritional needs from a vegetarian-type diet, if balanced carefully. They like cooked grains—corn pone, mush, oatmeal, bread—but whether you want to spend the time cooking for them is another question.

The way cats and dogs were fed on the traditional farm was efficient and nutritionally complete. At milking time, the pets got a pan of milk fresh from the cow, daily. When hogs, beef, chickens, and rabbits or whatever were butchered, they got parts of the carcass the farmer didn't want. They caught and ate rodents, to the farmer's great benefit. And there were always table scraps to paw through. When I was a kid, buying commercial pet food was unheard of on the farm, and we always seemed to have about a dozen cats and at least one dog around all the time.

If you live the traditional farm way, you won't have to worry about your pet cats and dogs unless you allow them to overpopulate. But if you don't raise your own milk and meat, you will have to spend considerable money on pet foods to supplement table scraps and the occasional rodent the pets might catch. In almost all cases, pets are fed too much and get fat, just like their masters and mistresses. Members of the Church of the Divine Pet also decree that barns are too cold for cats and dogs, which is a terrible mistake. Barn cats and dogs are healthier for being so, and without them you are going to get one of the garden farm's worst problem: rats and mice.

AFTERWORD

Not many writers get a chance to revise a book that they wrote thirty years earlier. There's an eeriness to it. I feel like Rip Van Winkle—like I fell asleep out in my corn patch and, when I woke up, things looked about like always, but it wasn't even the same century anymore.

The satisfying part of this eerie feeling is that much of what I said on the subject of small-scale grain raising thirty years ago is more current now than it was then. The pancake patch has come of age. If that sounds like a brag, I'll not apologize. To all those agribusiness experts who ridiculed my call to garden grains thirty years ago, I now draw myself up in pompous self-righteousness, stick out my tongue, and gloat as sickeningly as possible.

Seriously, though, I have little justification for gloating. Much of the credit goes to an editor and dear friend whom I worked under at Rodale Press, Jerry Goldstein. A book about garden grains was more his idea than mine. Although I was already doing most of the things I would write about in the book, I did not think very many other people were that crazy. I was raised up in the generation that decided farmers had to get big or get out, that local gristmills like the water-powered "Indian Mill" of my boyhood had faded away into ancient history (it's actually a museum now), and that local bakeries like Neumeisters' in my hometown were gone for good. One of the fond memories of my youth was fishing below the dam at Indian Mill and being in town about four o'clock in the afternoon when the bread was coming out of Neumeisters' ovens. That heavenly smell would float all over the village. Made me weak in the knees.

But what the heck. I was a struggling writer, and if Mr. Goldstein wanted a book about grains, I was the man for the job. I was a link not only between Wonder Bread and homemade buckwheat cakes, but also between vast commercial grain fields, where I worked as a young man, and the small homestead garden culture that was taking hold of society's imagination in the 1970s.

I was surprised by the good response to that first edition. Evidently I was not the only crazy person out there. There were all kinds of mavericks who were willing to grow wheat in the backyard, thresh it by hand, and bake really good bread with it or feed it to chickens for fresh eggs and southern-fried that would make Colonel Sanders weep with envy. And instead of being a fade-away fad, the book kept on attracting interest, so much so that when it sold out, a group of homesteaders in the Ozarks wrote a letter asking that it be put back into print. But editors were not convinced, nor was I, to tell the truth, that enough more copies would sell to make the printing profitable. At least I thought that way until used copies appeared this year on Amazon.com priced ridiculously at over a thousand dollars each. And, just my luck, I've only got one left.

So this new edition will have one thing going for it. Buying it will save you a nice little wad of cash over trying to find a good copy of the old edition. And, if I do say so myself, the revised version is considerably better.

Revising an old book that some people apparently treasure involves a problem for the author. Beyond correcting errors and deleting obsolete information that may not be so obsolete in the future, plus adding new relevant material, what more ought I to do? Because of age, I no longer raise pigs or milk a cow, for example, although I am sorely tempted to return to doing both. Should I leave the book in the voice of a younger man with more energy than good sense, or do I write as an older man who hopefully has learned a few things in thirty years?

So I tried to straddle the line between before Rip Van Winkle and after. The two Genes aren't really that much different anyway. I just have to be a little more careful with what I say nowadays because I can't run fast anymore.

Another problem kept bothering me as I revised the book, although it is not really a problem. Thirty years ago there was no Internet. So-called how-to books could fill a need just by passing on pure information. Today there is no pure information in the field of small-scale grain raising, or anything else, that is not done to death on the Internet. So I deleted some of the "facty" stuff, as I call it, that you can find easily at the click of a computer mouse. It actually made the book better, it seems to me, because that kind of information is so boring.

Then I did what the Internet can't do: I put more humor in the book, more anecdotes, and more of my own highly opinionated ideas. I'm fairly sure that's why people read books anyway.

An Illustrated Glossary of Grain Equipment and Terms

The main purpose of this glossary is to familiarize the reader with the tools of grain raising, under the (perhaps wrong) assumption that beginners in growing garden grains probably wouldn't recognize a corn "jobber" if it jabbed them in the face. I have eliminated many of the sources of tools that were listed in the first edition because so many of them are outdated; I kept only those I know to be still in existence. Actually, trying to tell people where to find older grain tools and farm machinery as well as new stuff for small growers is like telling someone where to find mushrooms. You just never know what's available in the next old barn you happen upon or the next Web site on the Internet. I am trying here mostly to tell you what to look for. The where is part of the joy of the hunt.

Planting

For at least a century, most of the corn in this country was planted with this type of planter. Old-timers more often called it a "job planter." Two variations on the same idea were built, but both of them (see next page) are basically glorified dibble sticks. The lower steel blade of the planter was jabbed into the ground and then, by moving the handles, three or four kernels of corn dropped from the planting box through a planting tube into the ground opened up by the blades. When the point of the planter was pulled out of the ground, the dirt fell around the seed.

Even after these planters became more or less obsolete, farmers used them to replant hills of corn skipped by the horse-drawn planter. As a child, one of my less welcome chores was to replant with the hand planter, which by that time we always called the "replanter." I still have one in good working order. It is very handy for interplanting. For example if you have a row of corn up and growing and want to plant pole beans beside the cornstalks, the

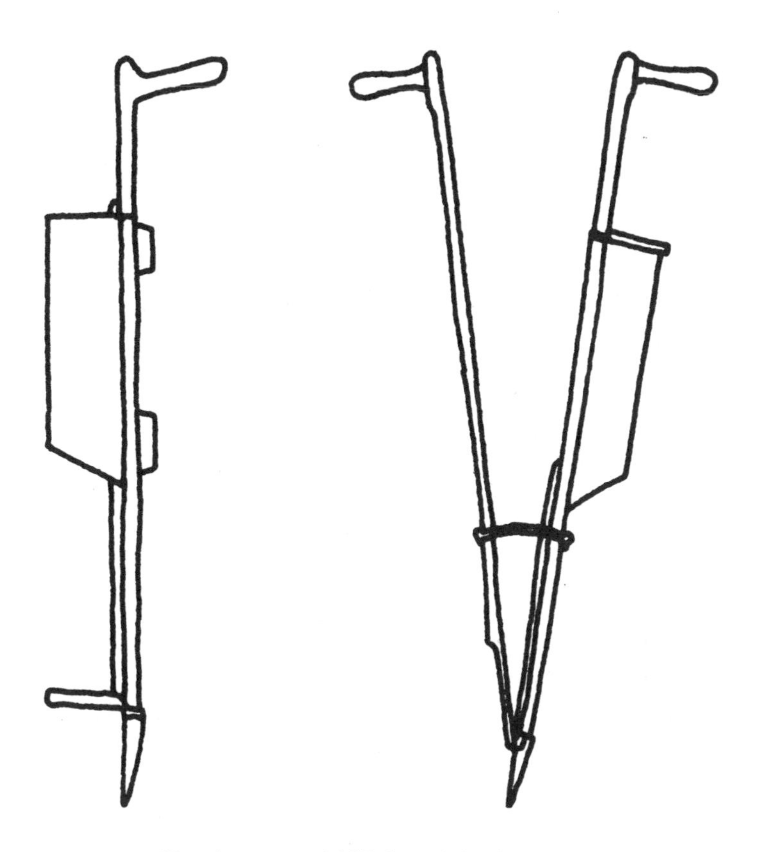

Hand-operated hill-drop jab planter

hand planter is just the ticket. You can walk along the row, plant the seeds where you will, and not disturb the soil around the already growing corn.

Fancy, newer models of the jab planter are still available, and are used in research plot planting. They are available from manufacturers and suppliers of the seed-industry equipment, such as Burrows Equipment Co., 1316 Sherman Ave., Evanston, Illinois, 60204. Older, used jab planters sell commonly at farm sales. You might pay $5 for one, or you might pay $35. You can never tell at an auction.

A picture is worth at least a thousand words to show you what the horn seeder looks like. It can be used to broadcast any seed you want to grow in a solid stand, and it gives, in the hands of someone who knows how to use it, a more even distribution of seed than you could do flinging it out with your arm and hand. Mostly I just wanted to include this tool for old time's sake. I don't know where

The horn seed sower

you can buy them today, although I bet they are available somewhere on the Internet.

A further refinement of the horn, the broadcast seeder hangs by a strap over your shoulder. Seed in the canvas bag (or metal hopper) flows through an adjustable hole in the bottom onto a fan powered by hand cranking. The fan spreads the seed evenly over the ground. Adjustments let you alter your seeding rate to suit the crop. You have to learn to walk and crank uniformly and then match the seeder's adjustments to your motions. Cyclone and Universal are the two brand names I'm aware of, available at or through most hardware stores and farm-tool catalogs. I wouldn't try to spread chemical fertilizer in broadcast seeders, especially the lighter, cheaper ones. The fertilizer will likely clog up the gears or rust any exposed metal parts.

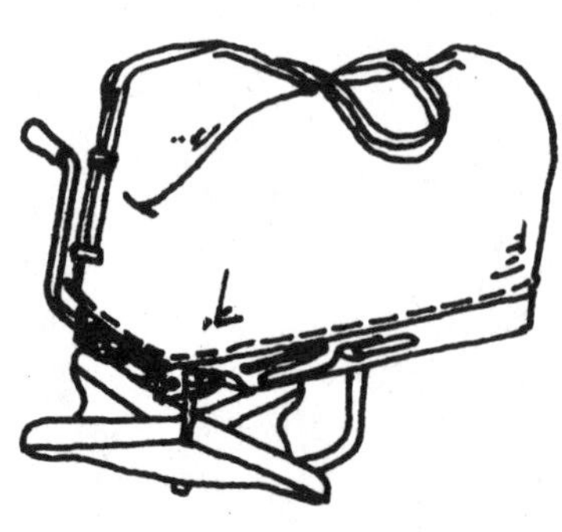

The horn seed sower

There is no more pleasant job on the farm than sowing a couple of acres of clover seed on a calm March morning with a hand broadcaster. The loudest sound you hear is the

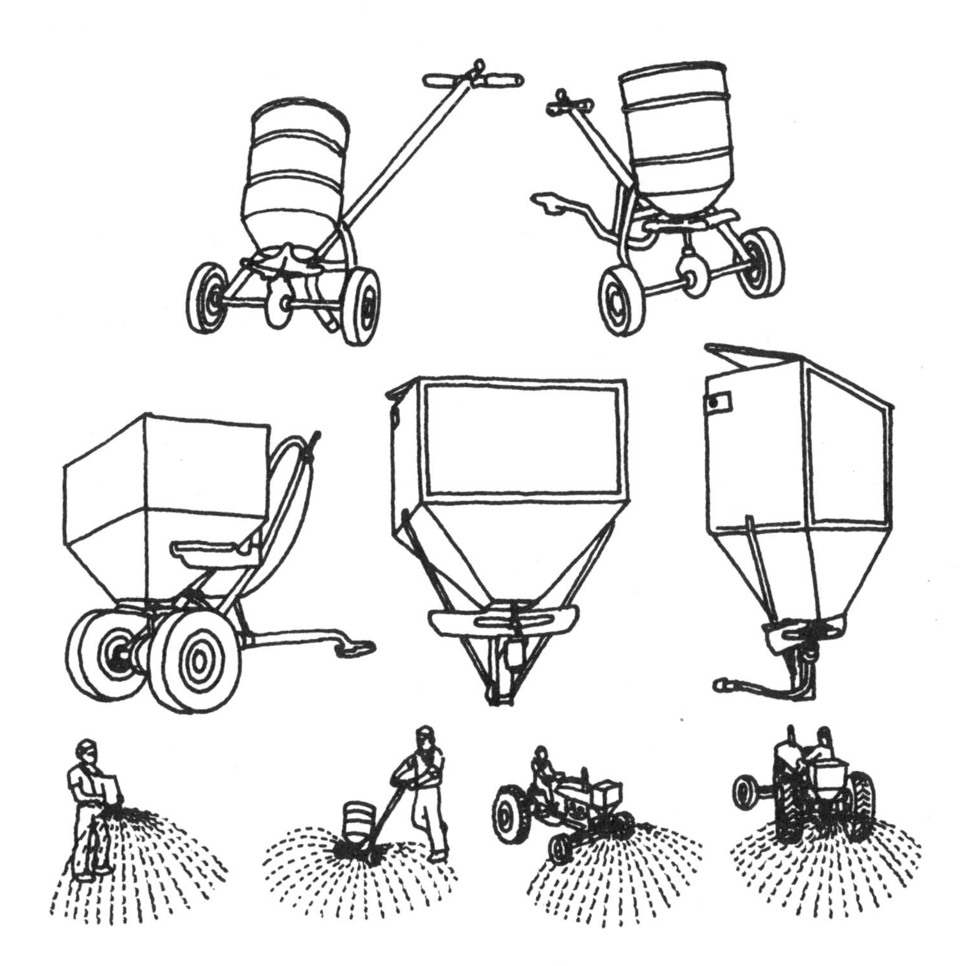

Hand-pushed row seeders

gentle whirr of the fan, and the birds singing. It is so quiet you can hear the tiny clover seeds strike the ground. And the machine becomes so much a part of your body, matching its action to the motion of your arms and legs, that you feel more like you are playing a musical instrument like a guitar than operating a machine. In fact, early broadcasters were powered by a bow and string rather than a crank, and the similarity to playing a violin, though crude, is hard to ignore.

Much larger broadcasters are powered by tractor power take-off, and are capable of sowing many acres a day. The "ultimate" in broadcasters today is the airplane, which farmers are using increasingly to plant grasses and small grains. That's a far, harsh cry from the lovely quiet mornings of hand sowing with a Cyclone broadcaster over your shoulder. Ain't technology wonderful?

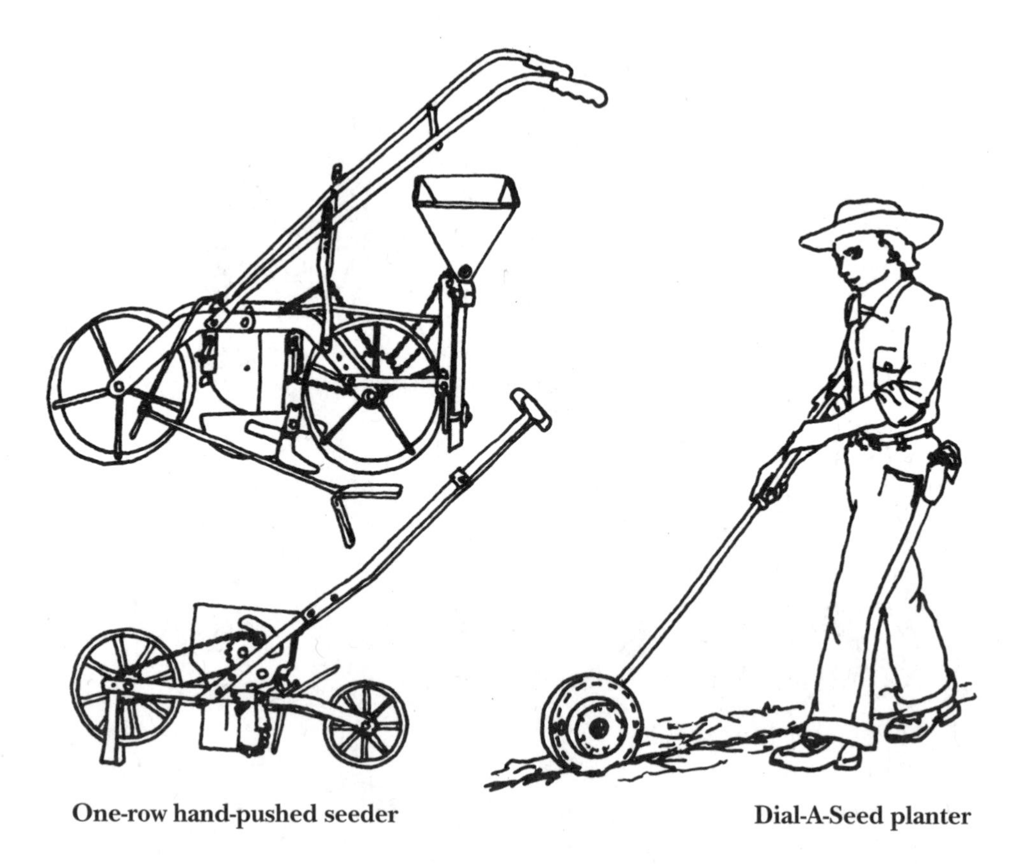

One-row hand-pushed seeder **Dial-A-Seed planter**

Small row planters mount the seedbox on a wheel (or wheels). The wheel not only carries the seedbox but activates a disk or plate, as it is called, in the bottom of the seed hopper. As the plate moves around, it allows the seeds to fall through the planting tube into the soil at regular, uniform spacings. In the simplest row seeder of all, for example the Plant-Rite and Dial-A-Seed planters, the wheel *is* the planting disk. The smallest two-wheel seeder with changeable disks for different planting intervals and seed sizes are regularly advertised and sold through catalogs and garden stores. A little larger and heavier planters are commonly available from manufacturers of small farm machinery.

All these planters can be purchased in models that will attach to garden tractors and can be pulled during the planting operation. Heavier commercial farm planters of great precision can now be purchased as single-row planters if you desire. The new planters, like those now available from John Deere, are called "unit" planters: each planting unit is powered by its own drive wheel, so you can use one alone or twenty together as big farmers do. You will

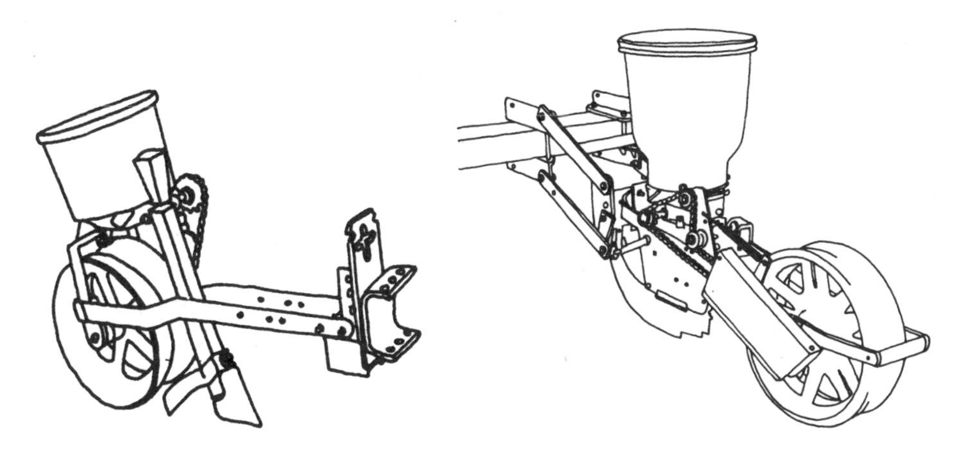

One-row seeders for mounting on a tractor

have to make your own hitch arrangement to attach one of these units to your garden tractor, but the small commercial grower may find it quite practical to do so since these planters will do, generally speaking, a better job of planting than the smaller garden planters.

Two-Row Planters

It may be practical with the unit planters mentioned above to go to a two-unit planter if you are row cropping more than four or five acres. The smaller Cole is sort of a unit planter; two units can be put together quite easily. Old, obsolete two-row planters originally for horses or smaller farm tractors also make fine tools for homesteaders with just a few acres to plant. But they are becoming

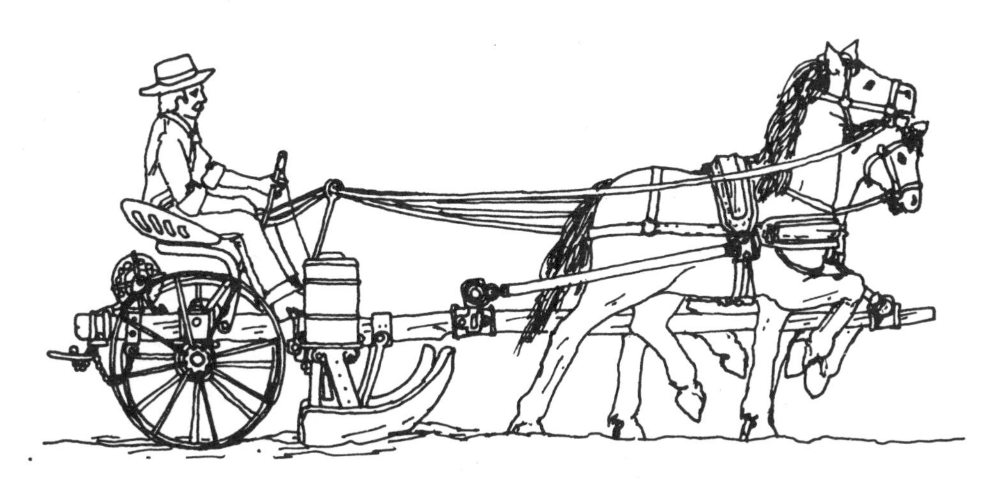

Horse-drawn two-row planter

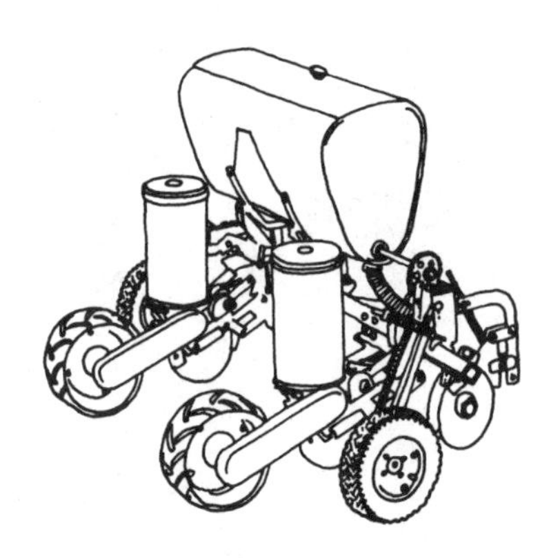

Two-unit planter

scarce now. Old four-row planters are now easier to find, but they are a little large for the typical homestead. However, you can modify one to a two-row model, or, if you are mechanically equipped, you can take a single unit from a four-row or two-row planter and convert it into a single-row seeder. It means, on four-row models especially, converting the press wheel that presses the dirt down on top of the planted seed into a drive wheel to turn the planting plates. Older four- and six-row planters do not have a drive wheel for each unit.

Normally, you can save quite a bit of money buying an old planter if it is still in running order. Be sure the planting plates come with it or can be purchased. This is not usually a problem now because very cheap plastic plates are made to fit most planter boxes. In fact, the plates are given away by seed-corn dealers when they sell a farmer his seed. The seed dealer wants a satisfied customer, and the plastic plates he has are made specifically for the size kernels he sells, and a different one for whatever plant population the farmer wants. But it is possible that on old and rare planters that you might chance to find, the plates would not be standard size, so you would have to make sure it has plates with it or know where to obtain them.

Finally, you need to be aware that modern large-scale farmers almost all have switched to sophisticated plateless planters. I will not even try to explain them here, but it is possible that the various new ways of planting seed will eventually "trickle down" to small-scale seeders.

Grain Drills

For planting seed in solid stands rather than in rows, the drill is used. (Small acreages can use small broadcasters much more efficiently, even taking into account the light disking or harrowing necessary to cover the seed.) The drill gives more precise planting depths than broadcasting and harrow covering, and will result in better germination, especially if dry weather follows planting. The drill puts seed into the ground more or less continuously rather

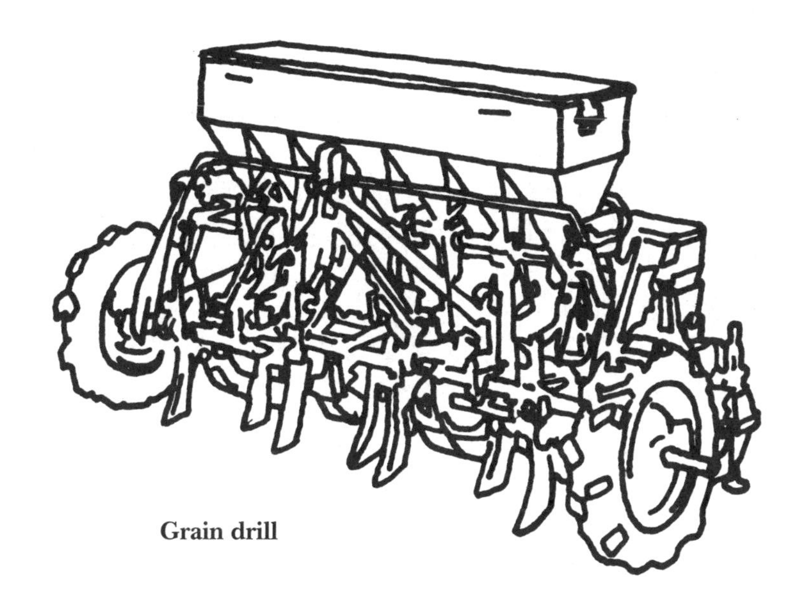
Grain drill

than a precise number of seeds at precise spacings the way a row seeder does. It is used for cereal grains and grasses, and sometimes for soybeans if a solid stand is desired. Essentially, the drill is a long planting box with openings every 6 inches or so, from which planting tubes lead to the disk "openers" that run in the ground at planting depth. The disks open a shallow trench for the seed to fall in, and close the trench on the seed after it falls. On old drills, a small length of chain drags behind each disk to help cover the seeds too. The seeds are actually planted in rows, just inches apart, and when the grain gets up to 6 inches high the plants grow together, giving the impression of a solid stand. The most common older drills plant a swath of 7 feet in one pass, and this is about the right size for a homestead.

Seeding rates can be regulated by controls under the seedbox. The seedbox may be divided into two compartments actually, and each side has its own control. Charts on the inside of the hinged box lids give instructions on how to set the adjustments for different seeding rates. On old drills you will have to experiment by putting a certain amount of grain in the box, then planting a known specified acre or portion thereof, and then see if the actual planting rate corresponds to what you have the drill set for. Compare and compensate accordingly. Those old cogs are worn and don't work as precisely as when the drill was new.

Many drills, even old ones, have fertilizer boxes just in front of the seedboxes. Unless well taken care of, the fertilizer boxes may

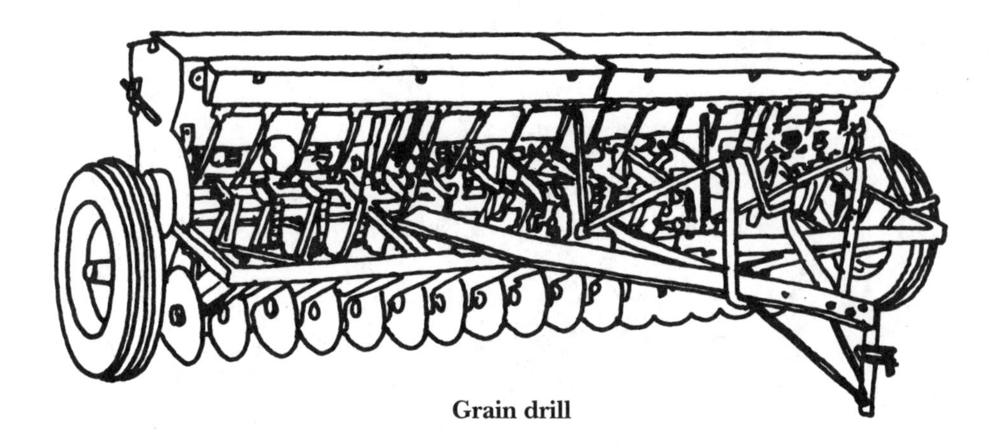

Grain drill

be badly corroded. Or completely ruined. This will not necessarily stop you from using the drill; you just won't be able to fertilize as you plant. In addition, some older and newer drills have special smaller boxes for clover seed so that when planting spring cereal crops like oats, the farmer can sow clover right along with it.

Seeding Rates

Charts make a book look impressive, so I'm going to include a couple against my better judgment. There is some general agreement on "correct" seeding rates, but they should not be looked upon as ironclad, infallible directives. Sometimes standard seeding rates are too high for a particular situation. (Rarely are they too low, because the rate makers are almost always the same people who are selling the seed.) Seed producers have told me, privately, that in many instances, lower seeding rates allow the plants to grow more seed heads and so give a yield approximate to the heavier, prescribed seeding rate at less cost. This is especially true of soybeans. Plant them thinly and they'll bush out more and throw more seed pods per plant. Plant wheat thinly in good ground and it will tiller more and throw more heads. Corn planted for an 18,000-plant-per-acre population will produce larger ears albeit fewer, perhaps, than at a rate of 26,000 plants per acre. Or produce more stalks with two ears. If you are farming organically without the intensively high fertilizer applications of chemical farmers, you will get a better crop at a 17,000 to 18,000 plant population than at the 22,000 to 28,000 rate. So use the following rate chart as a guide only. Don't be afraid to vary it up or down a little (especially down) if your experience, common sense, or necessity dictates doing so.

TABLE 7. SEEDING RATES

Crop	Seeding rate per acre in pounds	Weight per bushel in pounds
Alfalfa	15	60
Barley	100	48
Buckwheat	50	50
Cane (sweet) Sorghum	10	50
Red Clover	10	60
Sweet Clover	10	60
Ladino Clover	2	60
White Clover	2	60
Field Corn[1]	6–8[2]	56
Cowpeas	75	60
Flax	50	56
Golden Millet	45	50
Hungarian Millet	45	48
Japanese Millet	30	40
Oats	75	32
Field Peas	100	60
Rye	100	56
Rye Grass	25	24
Soybeans broadcast	100	60
row-seeded	40	
drilled solid	60–80	
Spelt	65	40
Sunflower	6	24
Wheat	100	60

1. On corn, better to use plant population as a guide. See below.
2. for organic growing

PLANT POPULATION PER ACRE

To determine the number of plants to the acre, multiply the distance between plants by the distance between rows in feet and divide that number into 43,560, the number of square feet in an acre. So, in corn rows 40 inches (3.3 feet) apart, and plants in the row 9 inches (0.75 feet) apart, the number you get from multiplying the two is 2.47. Dividing that into 43,560, you get a plant population of a little over 17,600. That, in my opinion, is about right for organically grown corn. If you are using open-pollinated seed, or seed more than one year old, or planting early in the season, you might want to plant closer to 19,000 population, to allow for kernels that don't germinate or don't grow for whatever reason. On my acre or two of corn, I sometimes plant at that rate or a little more and then thin out to 8 to 9 inches between plants where necessary.

On other grains, plant population is not an issue. Seed rates are reckoned only by weight. But remember, the bushel weights of grain given above are only for good, dry, premium grain. An actual bushel of wheat may weigh less than 60 pounds, indicating that some of the grains are light, chaffy, and won't germinate. If you plant such wheat, use a heavier seeding rate.

Weed Control

Hoe

This is still the best tool for weeding small plots of row crops, especially between plants in the row.

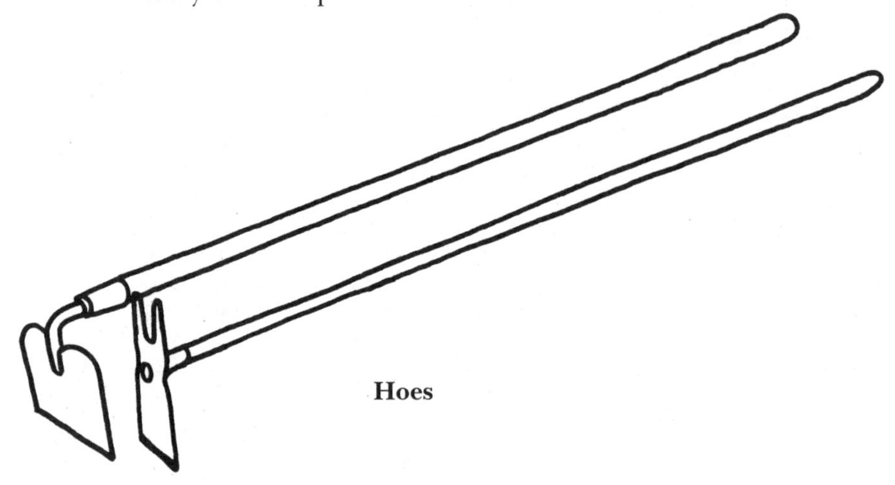

Hoes

Hand-Pushed Wheel Cultivators

A tool to cultivate between the rows faster than hoeing, the wheel hoe is practical on small to medium-sized gardens. Several brands are marketed and are widely available. The high-wheel types push more easily. A variety of tine and shovel blades are available for them.

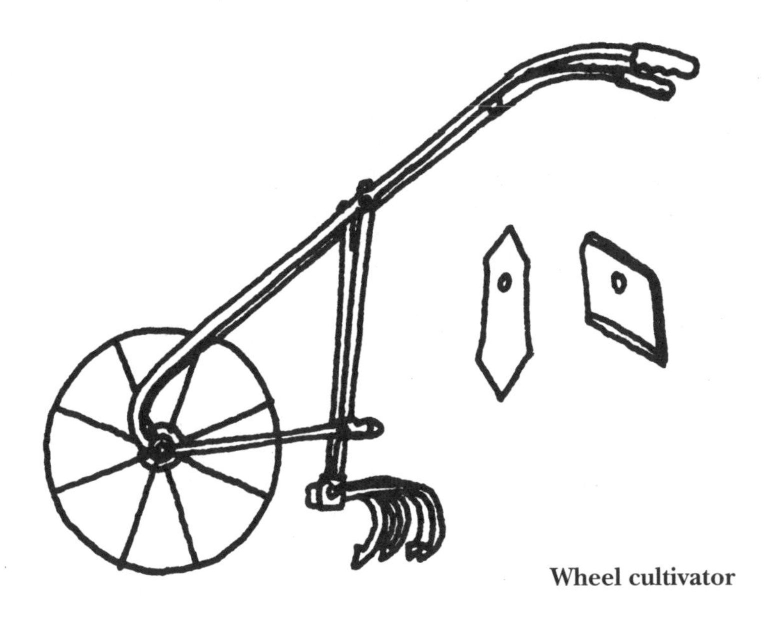

Wheel cultivator

ROTARY TILLER

Though thought of as mainly a soil preparation tool, the tiller makes a dandy weed cultivator too. Don't let it dig in more than three inches when cultivating. You need to be cultivating before the weeds barely start growing anyway, so a light touch is best, including weeding.

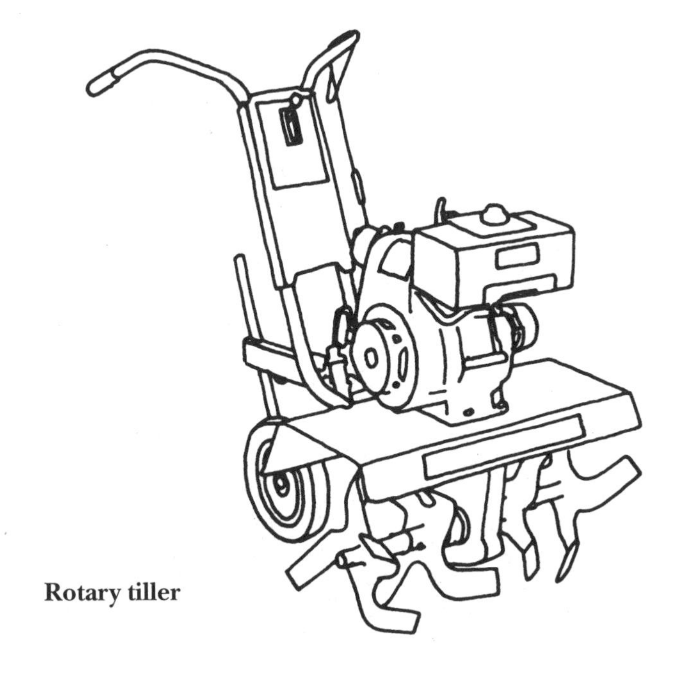

Rotary tiller

ROTARY HOE

This is a wheel (or drum) composed of slightly curved, spiked teeth that, when run through the soil at a fairly rapid speed, neatly toss very small, just-germinating weeds out of the ground. Small, hand-pushed types are sold at garden stores or through catalogs, but the really effective kinds are large, tractor-pulled models that until recently were standard on all farms. (Herbicides are making them obsolete, at least temporarily, but organic farmers like to use rotary hoes.)

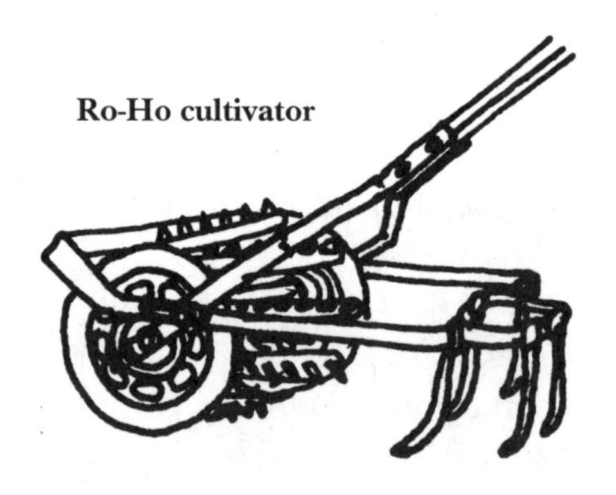
Ro-Ho cultivator

The rotary hoe is divided into sections, each of which will weed a swath about 3 feet wide in one pass. It pulls easily, and even the smallest farm tractor can tow two sections at the fairly rapid rate of speed necessary to do a good job, 8 to 10 mph. The rotary hoe should be run over corn and soybean crops about three days *after* planting, *before* the grain is up. Millions of weeds are sprouting, and the rotary hoe will kill many of them. The rotary hoe can be run over corn even after it is an inch above ground. You may throw out a stalk here or there, but no matter. (The harrow can also be used to weed in this manner, but it does not do as good a job as the rotary hoe.) This early preemergence cultivation is extremely important for organic growers. It is very important that corn get a head start on the weeds, and rotary hoeing insures that it does.

Rotary hoe

Row Cultivators for Garden Tractors

There are many kinds of row cultivators for garden tractors. Such lines of equipment are handled by various farm and garden machine retailers. Often, retailers in smaller towns do not keep such implements in stock, and you will not find them in their stores. Ask for them. Sometimes the dealer (or his assistant) is unaware of a particular type of small farm equipment until you ask him to look it up in one of his catalogs. Believe me, I've had that happen.

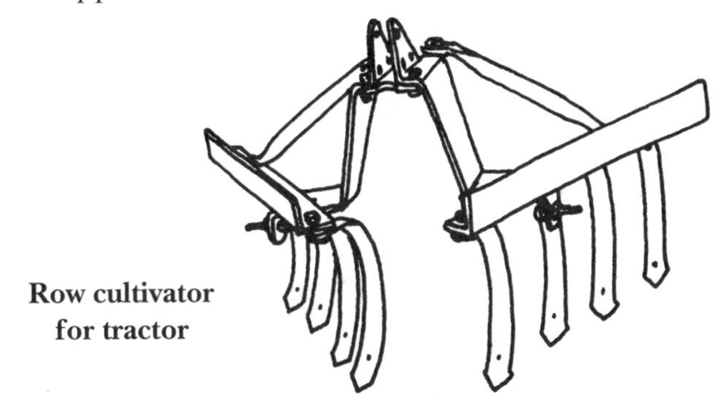

Row cultivator for tractor

Horse-Drawn Cultivators

Even if you don't have a horse, some old horse cultivators can be modified to pull behind your garden or farm tractor, and if so, can save you money. I've got a one-horse cultivator I bought at a farm sale for $12. The same or similar kind, if you purchase it new, obviously costs much more. I can pull this cultivator behind my garden tractor.

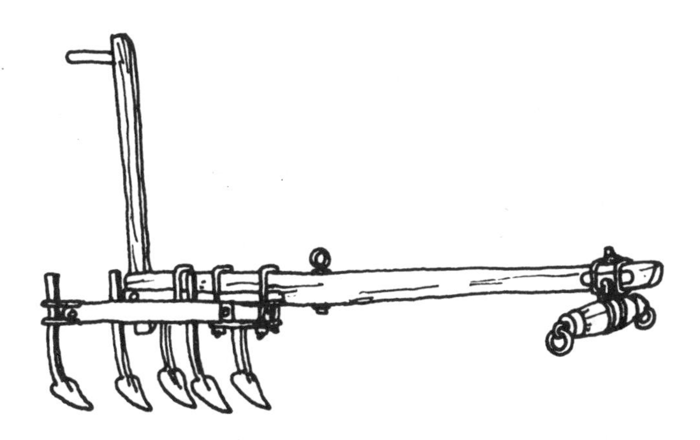

Horse-drawn cultivator

Farm Tractor-Mounted Cultivators

These cultivators, which weed two or more rows at a time, mount either to the front or rear (or both) of smaller farm tractors. Each tractor model has its own particular cultivator. The ones that mount on front or on both the front and rear are usually devils to attach and take off. Unless you can afford to keep an older tractor around with the cultivators attached to it all the time. I used to know a farmer who kept six older tractors just because he did not like to keep hitching and unhitching various tools from one or two tractors. Needless to say, he was a good mechanic.

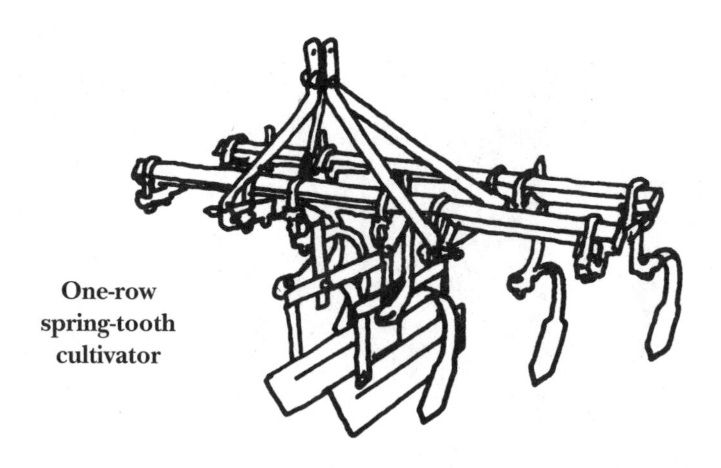

One-row spring-tooth cultivator

Today, almost all machinery connects to tractors by mounted three-point-hitch arrangements, which make it comparatively easy to back up to, attach, unattach, and drive to the field. Perfect for homesteaders is the old two-row cultivator made by Ferguson, Ferguson-Ford, and Ford companies for their little tractor of yesteryear. These cultivators will fit any three-point hitch, not just these particular tractors.

There are four-, six-, and eight-row rear-mounted cultivators, all somewhat too large for homesteads. But it might not be too ridiculous to buy an old four-row model cheap and cut the two outside cultivator gangs off, so you can use it wherever you planted with a two-row planter. For instance, I may do that, since my Allis-Chalmers tractor has its own hydraulic lift system different from the three-point hitch. Four-row cultivators were made to fit the Allis-Chalmers, and when I find one, I plan to buy it and cut it down to my size. The essential rule is: you can cultivate in one pass no more rows than the number of rows you can plant in one pass. Rows sown with a two-row planter can be cultivated with a one- or a two-row cultivator but not a four-row. A one-row planter can only be followed by a one-row cultivator.

Shields. All cultivators for tractors have shields that fasten to the cultivating shovels that ride nearest to the plant. The shields keep the dirt sliding off the side of the shovels from rolling onto small plants and burying them. You can even get shields for your rotary tiller, if you want to use it to cultivate close to plants.

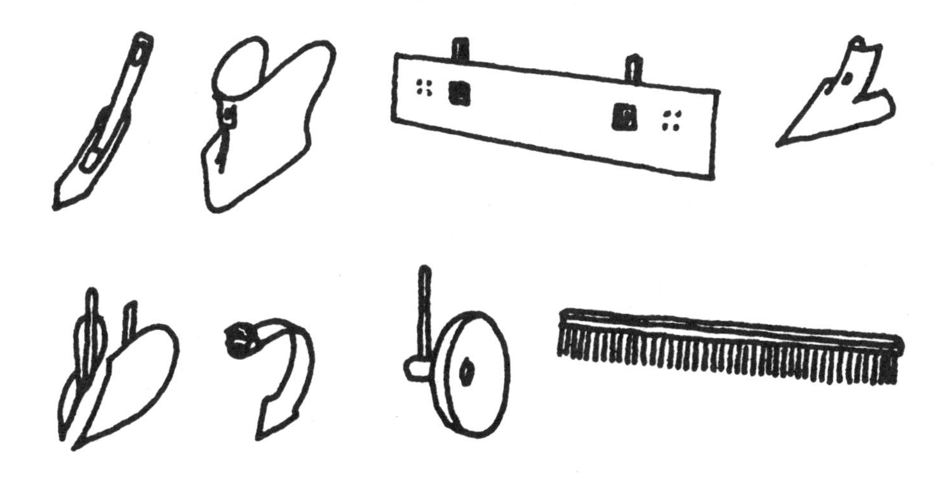

Various cultivating shovels and blades

Shovel and tine cultivating blades. Blades on cultivators come in many sizes and shapes. Shovel blades may be narrow and long, wide and short, diggers or sweeps. You can spend days fiddling around with various kinds to see which dig out weeds best and leave the least amount of furrow behind. I'm not about to tell you where and when to use sweeps because I'm not convinced that it matters. What does matter is that when using gangs of cultivators, each with its own depth adjustment, be sure they are all set to dig at approximately the same level and none too deep. If it's too deep, a shovel can harm roots or leave a big rut in the middle of the rows. If too shallow, it won't get the weeds. Shields should be removed after plants are six inches tall or thereabouts. Then the dirt rolling off the shovels will cover small weeds growing in the row and save much hand weeding.

Harvesting

Corn Knife

A corn knife is a long swordlike blade for cutting cornstalks by hand. Another version has a long wooden handle with a very short sickle blade at the end of it.

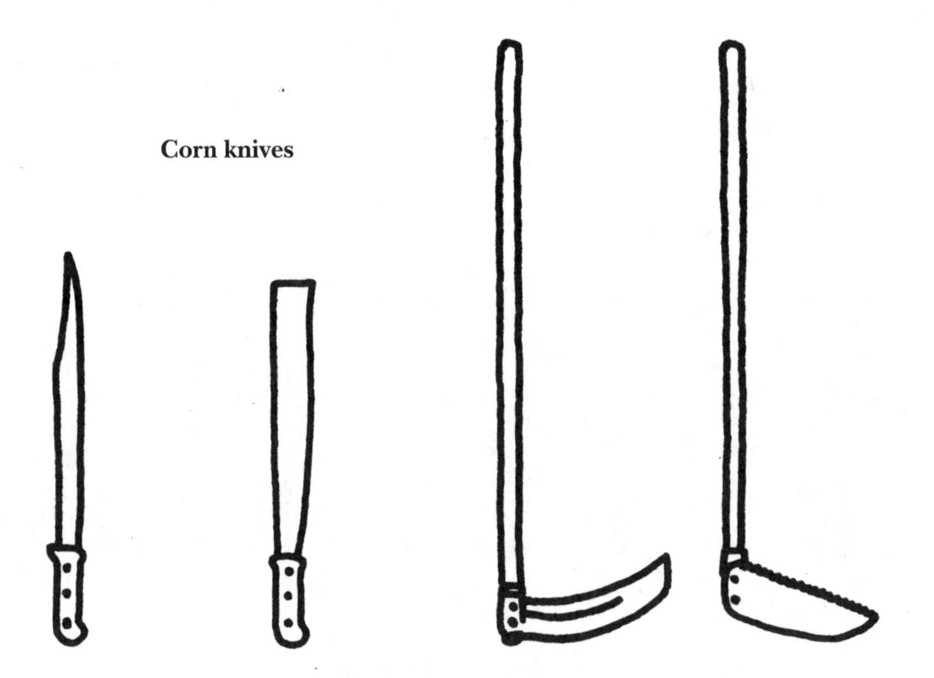

Corn knives

Husking Peg

Any of a variety of handheld tools called husking pegs are used to strip husks from ears of corn in the act of husking the ear. Some are very simple wood or steel pegs with a leather thong to hold the peg to the husker's hand. Others are steel hooks attached to a piece of leather that straps around the hand. I have one made entirely of brass that clamps to the fingers in a way that reminds me of brass knuckles.

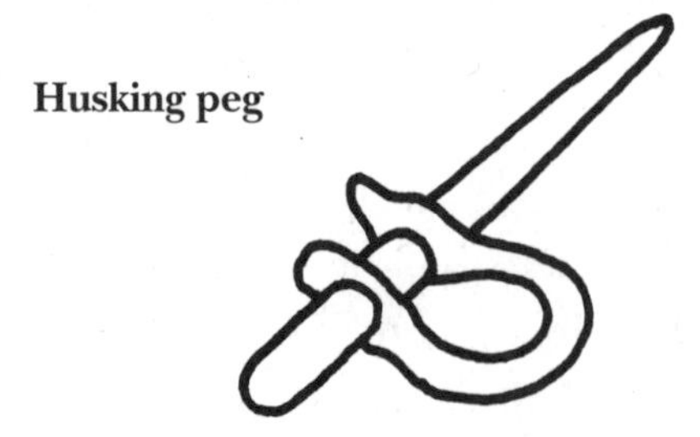

Husking peg

Corn Binder

Rarely used anymore, a corn binder could still be handy on a homestead, if you can find one in running order. The binder cuts the stalks of corn with a small reciprocating sickle bar, then bunches and binds the stalks into bundles with an automatic tier. They were made for both horses and tractors, and some were powered by the tractor's power take-off. I have very intimate knowledge of the latter, since one such machine nearly killed me, and on another occasion nearly succeeded in breaking open my rib cage. Be careful around machinery; it is dangerous. Be especially careful to keep shields over power take-off shafts *in place.* I didn't. Another picture out of my past I will never forget: my mother singing and driving the horses and binder down the corn rows, while I rode one of the horses and my baby brother slept in a wooden box bolted to the binder tongue.

Corn binder

Grain Sickle

A grain sickle is a short-handled, long-bladed sickle used for centuries to cut cereal grains planted in solid stands, and still used in some parts of the world.

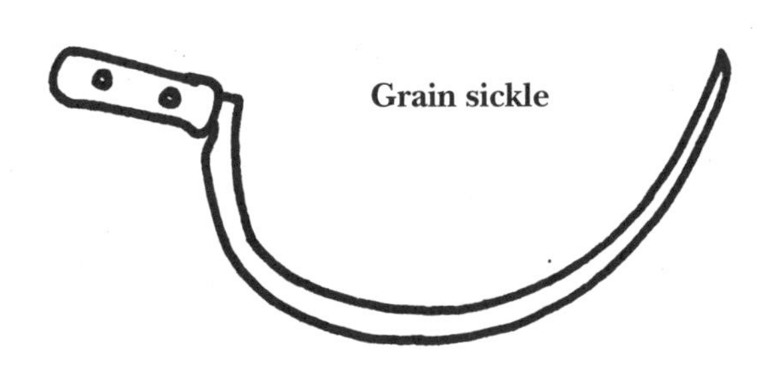
Grain sickle

Corn Picker

The commonly accepted colloquial name for the machine that replaced binders and the whole old method of harvesting corn is a corn picker. The picker did not (does not) cut the corn stalk off, but merely strips the ear off as the machine moves down the row, husks the ear, and drops it into a wagon from a conveyor chain into a trailing wagon. Old one-row pickers are often available at farm sales or may be rusting away in old barns. They are offered for sale regularly in the want ads of rural newspapers. Tractor-mounted pickers came along later. Their advantage is that the operator can pick any twin row in the field without knocking any corn down, as is the case with a pull-type picker. But they are difficult to attach to the tractor, and I would advise a small-scale corn grower to get a pull-type picker. Husk the outer rows that you might run over with the tractor by hand.

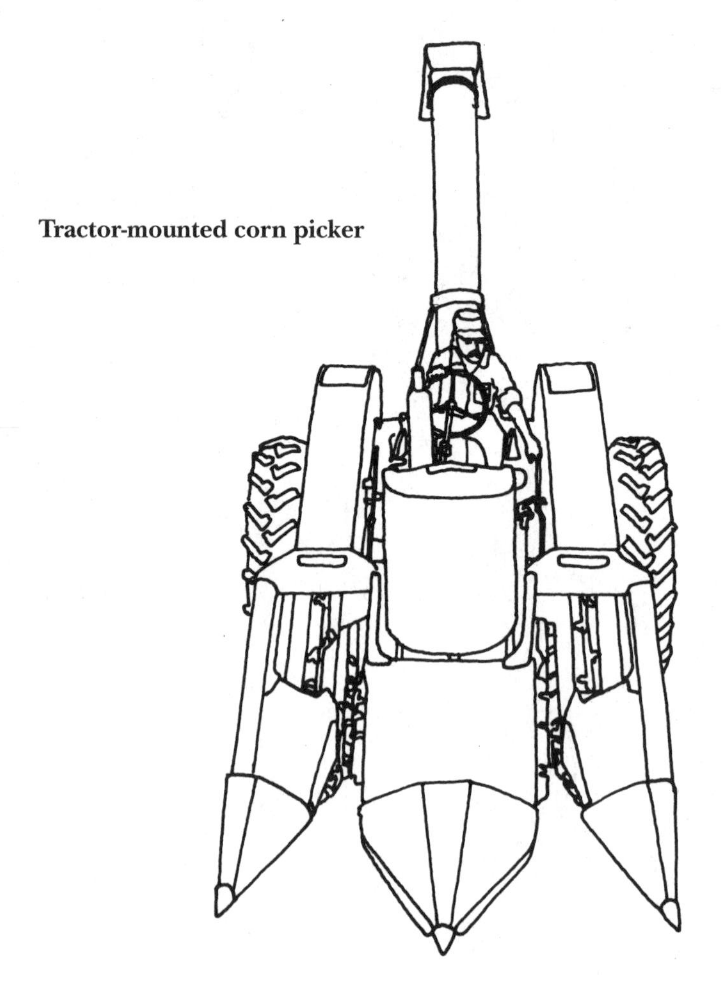

Tractor-mounted corn picker

Corn Combine

Now simply referred to as a combine, since the same machine with a different header and screens harvests the other grains too, the picker-sheller is replacing the picker on most commercial farms. In addition to doing everything the picker does, this modern machine shells the corn too. For the all-grain farmer, this is one modern machine that really does improve the soil organically, since even the cobs are left in the field along with the stalks, all to rot back into humus. The machines are terribly expensive and out of the question on small farms, but if you hire someone to harvest your corn, which is often the most practical thing for you to do, this is the method and machine that will be used. Don't think that by leaving the corn stalks in the field that they are necessarily wasted. Livestock and horses can winter on the stalks.

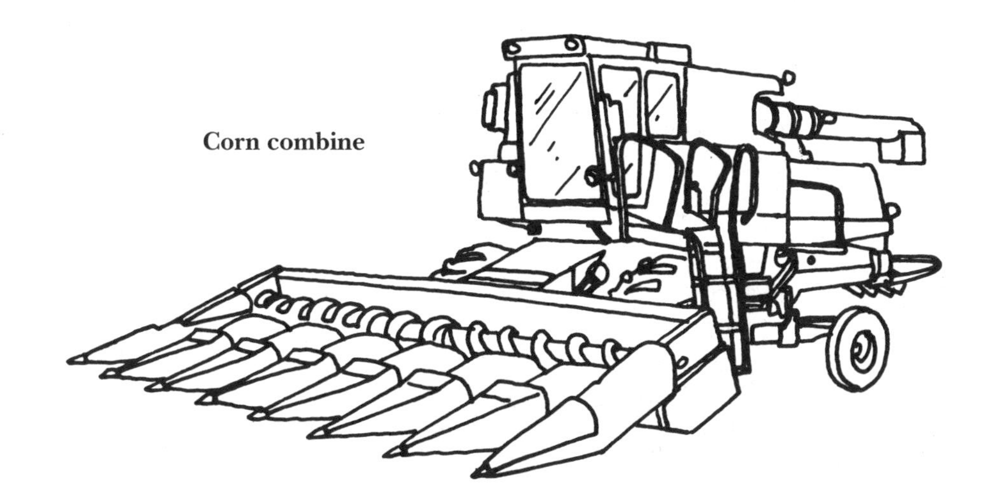

Corn combine

Grain Flail

Connect two sticks by a leather thong and you have a grain flail for beating grain from seed heads. A plastic toy bat works pretty well too, and a friend of mine says a piece of rubber hose works even better. Though obsolete, flails are practical enough for small garden plots of grain raised for your own consumption.

Grain flail

Grain Cradle

This is a special scythe with wooden fingers above the blade to catch the cut stalks and bunch them for easy tying. The cradle made the sickle obsolete in many parts of the world beginning about the seventeenth century, but, curiously, not everyone adopted it. There was probably as much grain cut in this country in early days with a sickle as with a cradle. The latter was a good deal faster. Use of the cradle on small farms in the South was reported by the USDA as late as the 1950s, believe it or not, and I suppose a few are still in use for harvesting small plots. I'd use one if I had it.

Scythe

Meant for cutting weeds and grass by hand, the scythe can also be used to harvest all small grains like wheat, oats, and rye in a garden-grain situation.

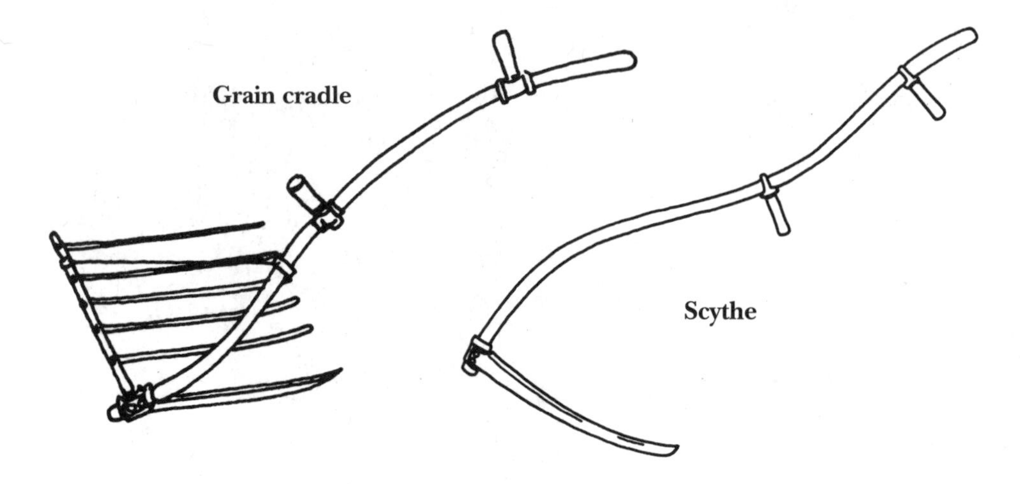

Grain Binder

A tool similar to the corn binder is the grain binder. It automatically cuts and binds small grains into bundles for shocking. Although usually pulled by horses, some late models were adapted to tractors. The grain binder is all but out of use now, except by farmers of particular religious sects who keep up the old ways (because they are a lot smarter than we think they are). I used one as late as 1954 in Minnesota and know they are practical for small fields even today, *if* you have a machine thresher to go with it.

Winnower (or Separator)

Any system that separates grain from chaff, usually (or always) by wind power, is a winnower. Grain can be winnowed by letting it fall from one bucket to another in a stiff breeze. Any type of fan works too. I use a big window fan. Mechanical threshers have built-in fans to blow away the chaff.

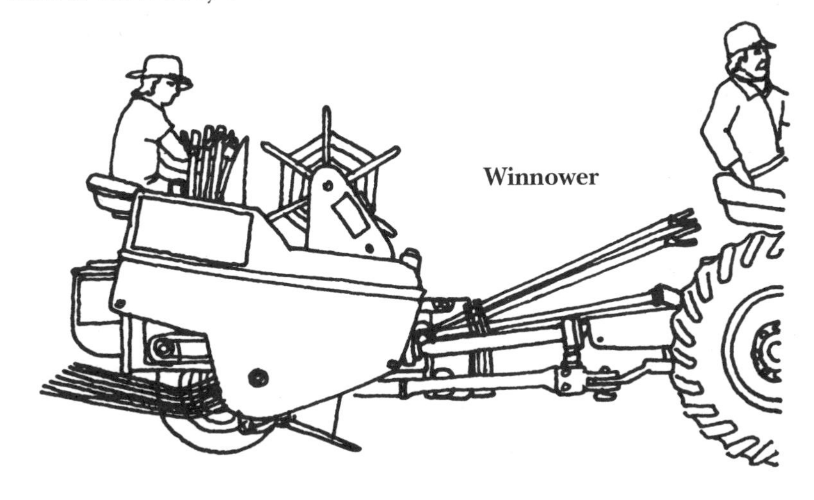

Winnower

Grain Thresher

Any machine that combines the beating action of the flail and the winnowing action of the separator mechanically is a grain thresher. The huge stationary threshers on yesterday's farms also blew the straw into a large stack for winter bedding. These old threshers are seldom used anymore except in historical exhibitions.

Small threshers are still made and sold by the seed-equipment industry. They are used to thresh seed in research and experimental plots. (See chapter 3.)

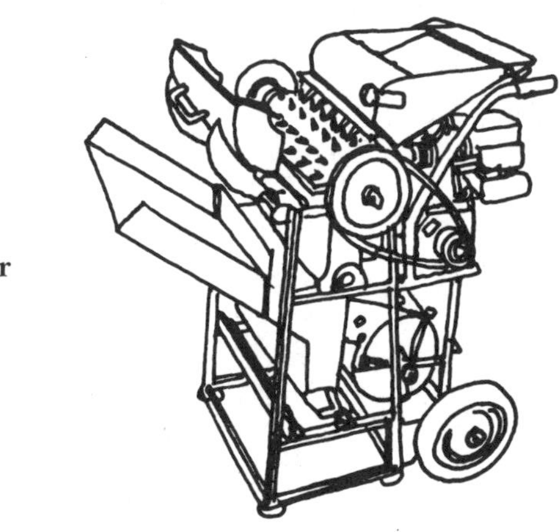

Grain thresher

Grain Combines

This machine combines the work of both the binder and the thresher, hence the name, "combine." The minimum size now made for commercial farms cuts a swath over 10 feet wide. Older, smaller combines varied from 3- to 7-foot cutting widths, and would be very practical today for homesteaders if you can find one in good shape. Even if you have to put $400 of necessary new parts into an old combine that you've bought at a price of $100, it is still a good bargain.

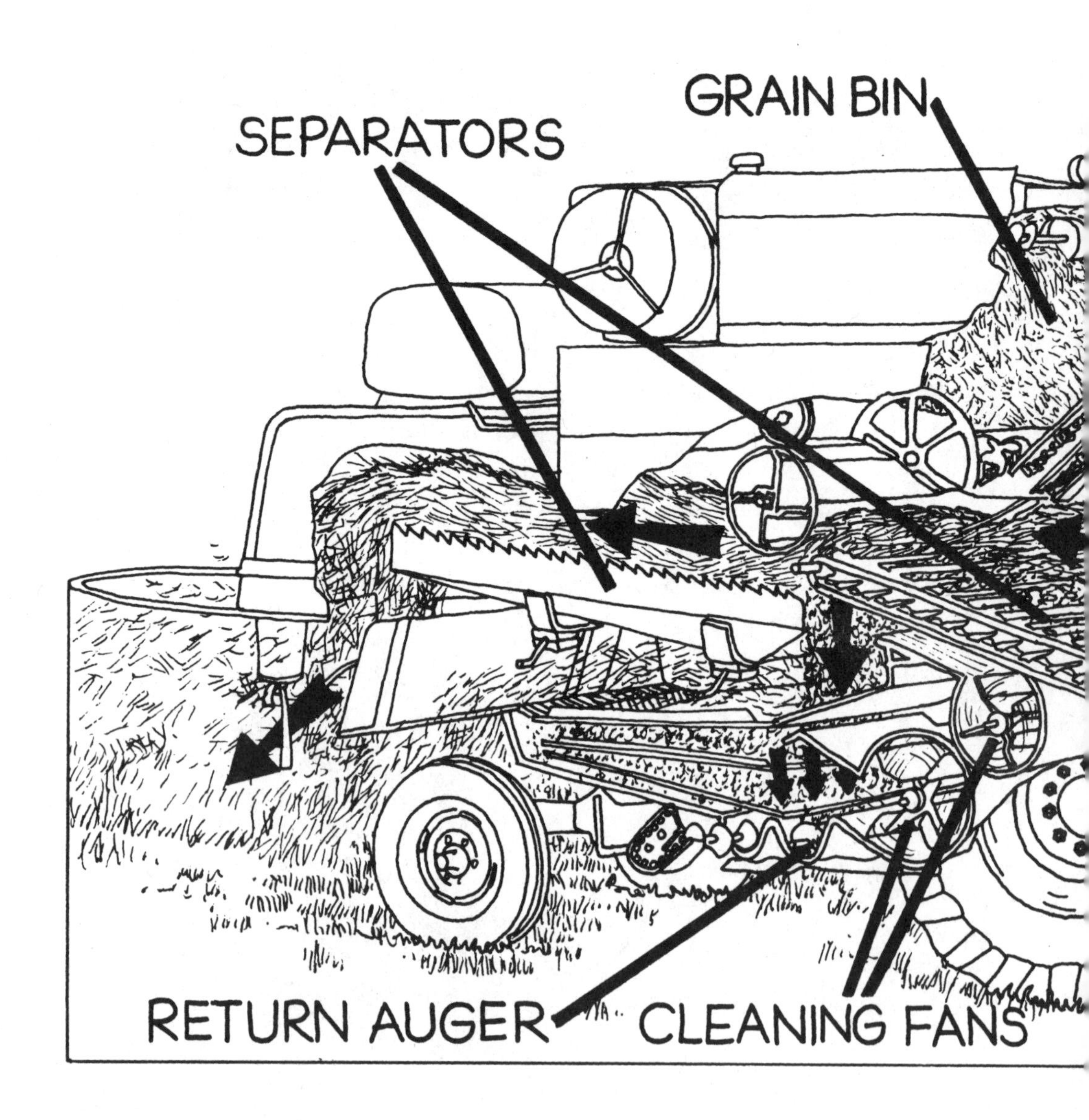

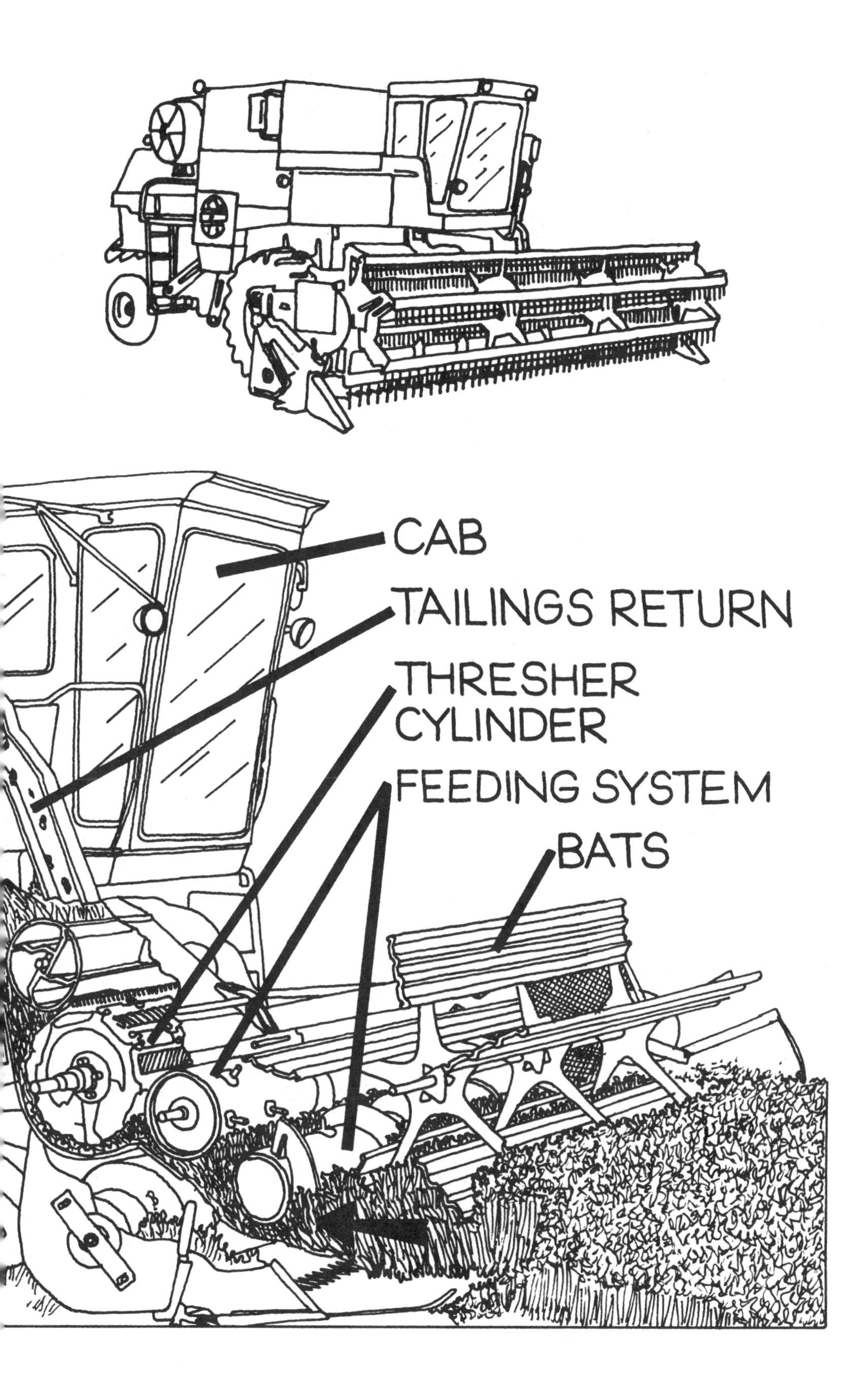
CAB
TAILINGS RETURN
THRESHER
CYLINDER
FEEDING SYSTEM
BATS

Very small combines are made for research plot work; universities and other institutions of sophisticated research are about the only customers for the expensive little things.

Of the smaller, now more or less obsolete farm combines, I am partial to the Allis-Chalmers All-Crop because I lived with one so long I got to know its every quirk and foible, which were considerable. You can still find them in good shape—a friend of mine just found one and bought it very reasonably. But unless you are quite mechanically minded and can get one for next to nothing, I think you would be better off to hire a custom combiner to harvest your crop.

CORN SHELLERS

Most homesteaders will harvest their corn as ear corn and so might need to shell it for feeding to animals or themselves. Ground feed for chickens and hogs requires shelled corn, not to mention meal for human use. (For cows, the whole ear, cob and all, can be ground into the feed.) For shelling, every size implement has been made at one time or another, from a very small handheld sheller handy for shelling out your popcorn, to very large tractor-powered shellers that can shell out a whole crib full of corn in an afternoon. In between are hand-cranked shellers and small motor-driven models. Hand-cranked shellers are still available from Nasco, already mentioned, and from Lehman's Non-Electric catalog, P.O. Box 41, Kidron, Ohio, 44636. In fact, between these two catalogs you can find nearly all old-timey hand tools.

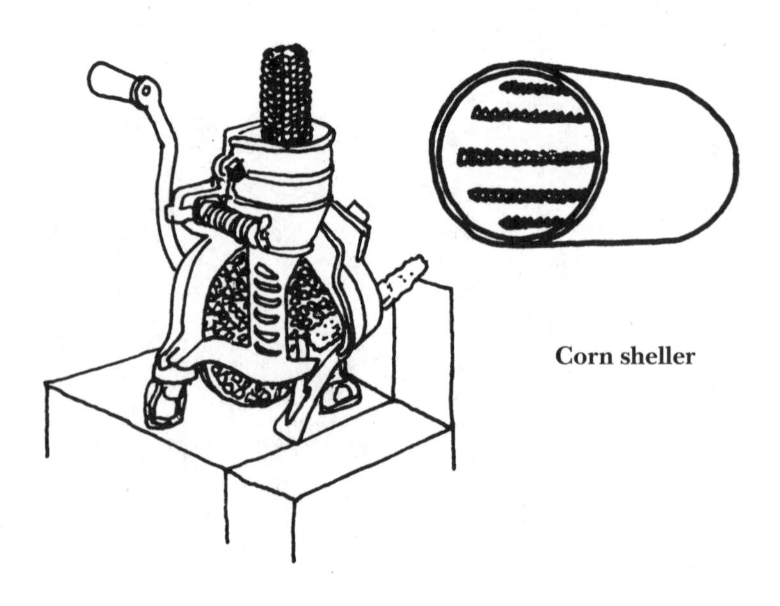

Corn sheller

But neither catalog offers the old heavier ones you can still buy at farm sales that have their own wooden frame. These stand about three feet tall, and the crank has a flywheel that makes turning much easier on your arm.

Seed Cleaners

These machines were standard equipment on farms when farmers saved their own seed for planting. They had to get the weed seeds out, especially out of clover. Most farmers buy their seed each year now, a practice homesteaders might find reason to avoid at some real savings of money. Modern combines clean out the weed seed and chaff fairly well, older ones not so well, and, in either case, not well enough either for seed or for grinding for table use.

At almost every farm sale I attend, an old seed cleaner usually sells for around $50. New ones cost from about $150 on up, available through seed equipment industry sources. The cleaner "shakes" the grain across a series of screens where anything larger or smaller, lighter or heavier, is sifted or winnowed out. Some are hand-cranked or converted to a motor. Quite fascinating to operate, I think. Pour dirty, chaffy clover seed in one end and it comes out the other as so many million clean and lovely little yellow-

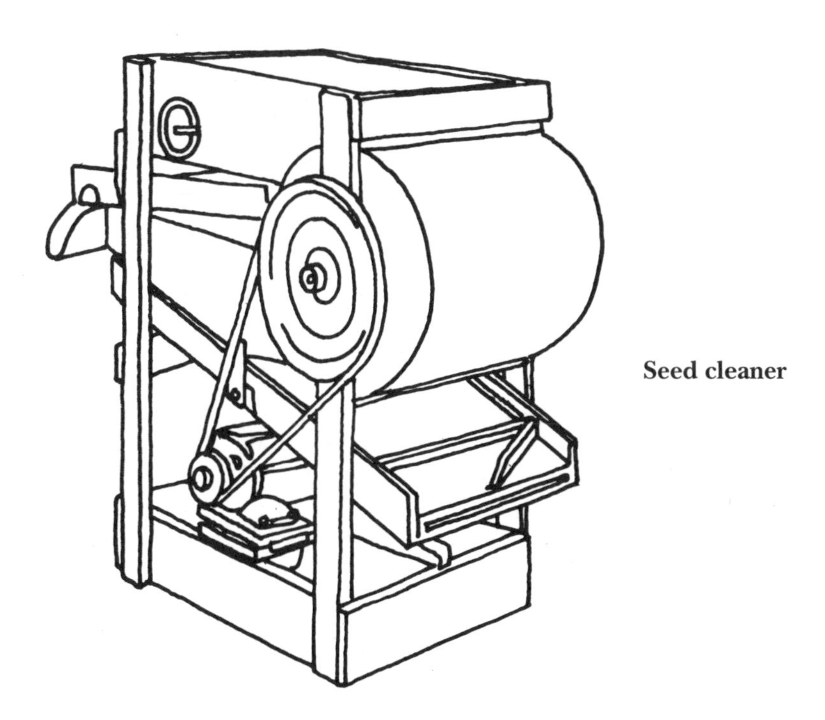

Seed cleaner

reddish-gray pearls, while dirt, chaff, and weed seeds dribble out of separate spouts.

Bushel Basket

No farm worthy of the name ought to be without a metal bushel basket. Next to the pitchfork, no tool was more common in days gone by. You will find one most useful.

Bushel basket

Grain Conveyor

You may on occasion hear this term and wonder. It refers to the long, metal troughs with conveyor belts or augers in them, the whole mounted on two wheels so that it is mobile. They are used to move grain from one bin (or wagon) to another. Grain is funneled or shoveled into the hopper at one end, and is carried upward in the trough or pipe to the opening or window in the crib or

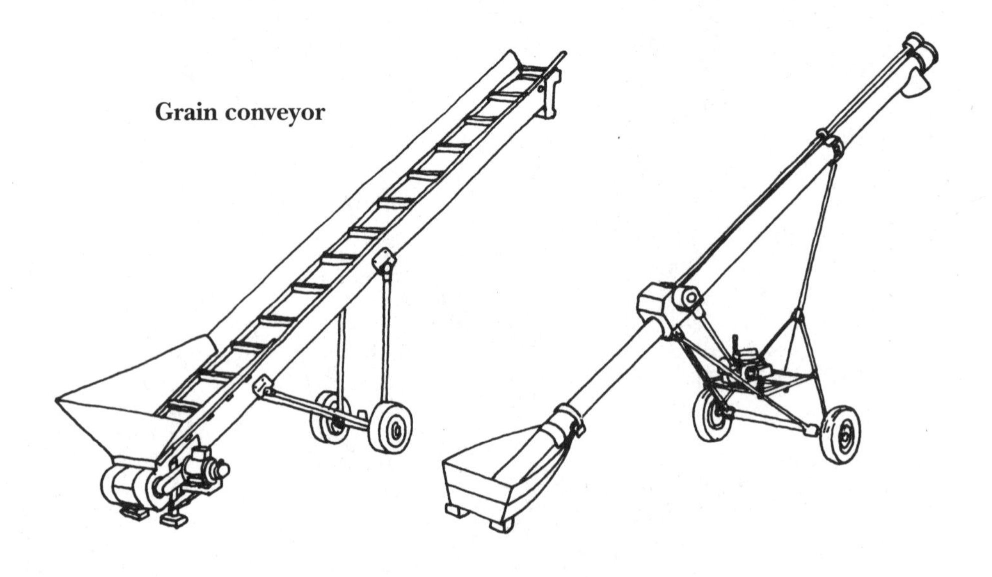

Grain conveyor

bin being filled. Obviously, one of these is necessary on any farm where the crib or bin is too high to shovel into directly.

CORN DRAG
A fork with tines bent at right angles to the handle is a corn drag. It is for pulling corn piled in a crib down to the exit or door. Ear corn is notorious for sticking to its pile and rarely will roll down by gravity to the floor where you can scoop it into a shovel.

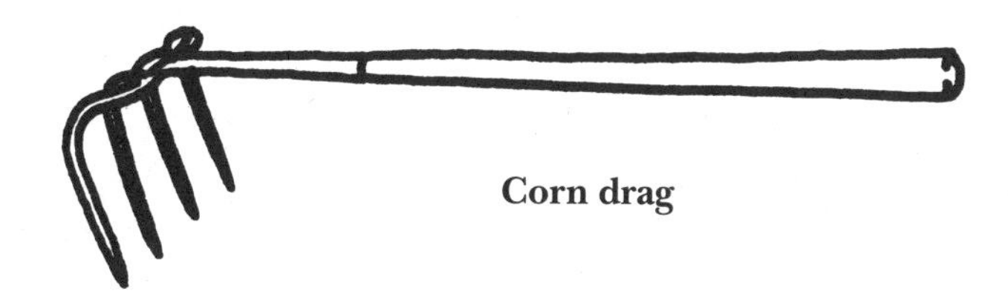

Corn drag

SCOOP SHOVEL
Every homestead needs a scoop shovel if grain is to be handled by hand. Aluminum scoops are *much* lighter and therefore more desirable than steel ones.

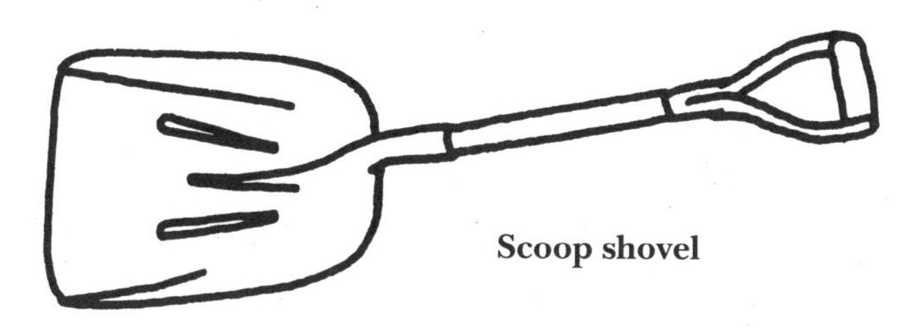

Scoop shovel

Processing and Storage

RICE HULLERS AND SHELLERS
Machines for processing rough rice to a condition where it can be used for food all seem to be of the large commercial size in this country. Smaller equipment is made in profusion in Japan and elsewhere. CeCoCo, Chuo Boeki Goshi Kaisha, P.O. Box 8, Ibaraki City, Osaka, Japan, puts out a most interesting catalog for the homesteader, but the cost of its tools has been prohibitive.

Rice huller

Barley pearler

Barley Pearler (Huller)

Both hand-powered and motor-driven barley hullers are available from the seed-equipment industry (see chapter 7). A machine that dehulls barley ought to work on oats and buckwheat too. Look on the Internet for the latest developments, as more and more people get interested in homegrown grains.

Sorghum press

Sorghum (or Cane Sugar) Press

You might, if you are very lucky, locate an old sorghum press rusting away in a southern barn. Otherwise, you will have to find someone who operates a press and mill. I have stated flatly in previous years that very few if any sorghum mills still operate in northern Ohio, only to find that I was very wrong. I now know two within easy driving distance of my farm, so I imagine this is true of most of the Midwest. And of course sorghum mills are common in the South. The Japanese company mentioned above (CeCoCo) makes a hand-cranked press or squeezer that I would love to own.

Crib

The term usually used for the building where ear corn is stored in is a crib. See examples in chapter 2 and below.

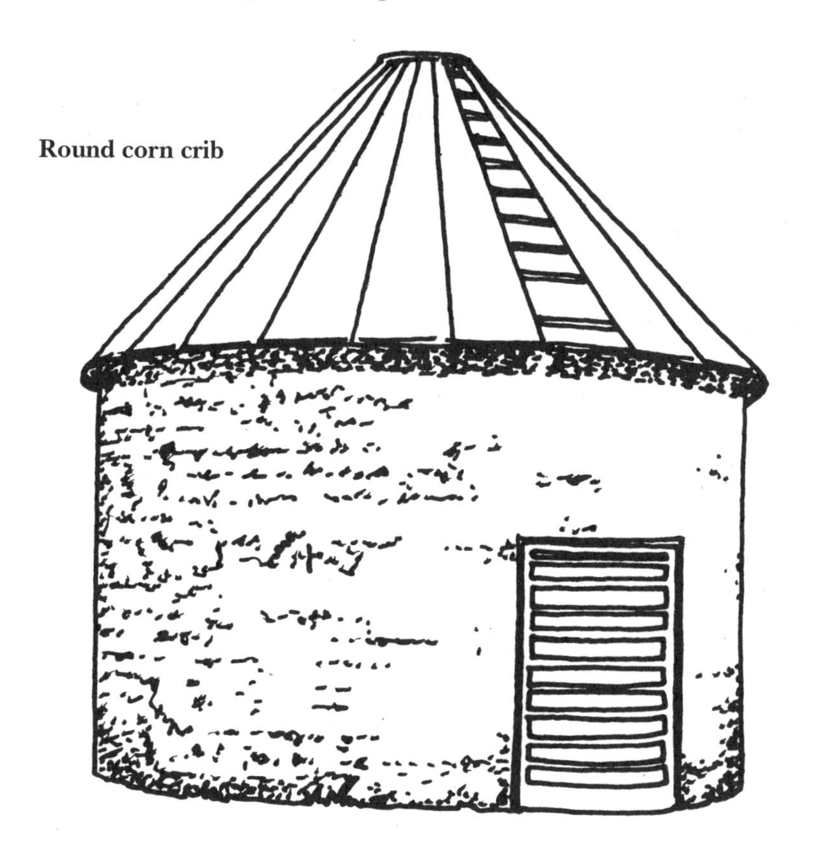

Diatomaceous Earth

Diatomaceous earth is an organic insecticide that has been found to be effective in controlling insect infestation in stored grain, especially weevils in wheat. The material is a finely ground talclike powder made from the fossilized shells of tiny one-celled plants called diatoms. See chapter 2.

Neem Seed-Kernel Powder

This is a substance used to prevent weevils from infesting stored grain. It is used commonly in India. The leaves of the neem tree are also used for the same purpose. Scientific tests of this traditional practice have indicated that neem seed powder is effective against insects that feed externally on stored grain, but is not so effective on borer-type insects.

Bin

Any storage room for small grains, beans, or shelled corn is a bin. The difference between a crib and a bin is that the former has openings in the walls for air circulation to complete the drying of the corn. Bins for grains other than ear corn are of tight construction, so that no kernel can leak out and no rain whatsoever can get in. Grain must be dried (13 percent moisture or lower) for bin storage unless special bin-dryers are present to blow air through the grain and dry it. See chapter 3 and the illustration below.

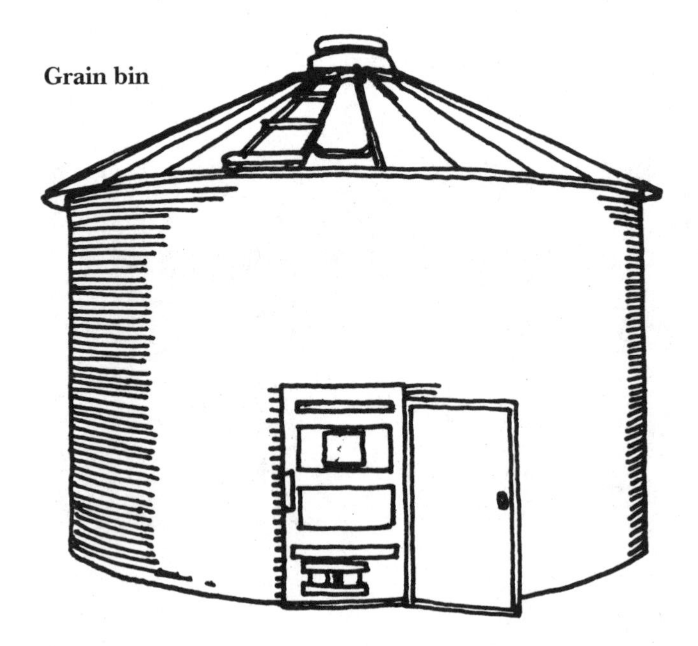
Grain bin

Dry Ice for Insect Control in Stored Grain

I won't vouch for either the safety or effectiveness of dry-ice treatment for insect control, as I have no personal experience with it. People who use this method, however, say it should be done this way: line a metal can (a clean trash can works fine) with two plastic bags, one inside the other. Pour about an inch of grain in the bag, put a piece of dry ice approximately the size of a walnut on the grain, then pour the bag nearly full of grain. Tie the top of the bags loosely so air can escape as the carbon dioxide gas from the ice rises. Leave overnight, in which time the piece of dry ice will evaporate. Then tie the bags airtight and put a tight-fitting lid on the can. Store in a cool dry place. The grain should be very dry for this type of storage. Otherwise it will mold inside those bags. Now, thirty years after I first wrote this book, I don't think this information is all that important or practical for small-scale grain growers. But it is something interesting to know.

Cold Storage

Cold storage is the easiest and safest way to keep grains from becoming infested with insects. It doesn't take much space to store a goodly supply of wheat in a freezer or refrigerator. Some folks keep secondhand refrigerators in the basement for drinks. Why not for grains? Also, here's a note of cheer. On corn, oats, barley, sorghum, buckwheat, and soybeans, which for many years I have kept around my house and barn, I have never had problems with weevils or bugs. Only on wheat and dried beans, and then only rarely. I believe the secret is to keep grain in bug-tight, or at least rodent-tight containers and use it up within six to nine months after harvesting, *and never keep it more than a year.*

Feed Grinders

Feed grinders are any of various machines designed to reduce grains to meal for easier digesting by animals or humans. There are three basic kinds: (1) burr mill with either stone or steel burrs; (2) roller mill; and (3) hammer mill.

Burr Mill

This is the old-time gristmill of the ages, driven by water power, wind power, or mule power. On farms, some steel-burr tractor-powered mills are still in use and in fact can be bought new. On all these mills, the basic principle—crushing grain between two heavy wheels, one stationary and the other revolving—is the same. New burr mills are available from or through farm-implement dealers, who will probably try to talk you into buying a hammer mill or roller mill instead. If, however, you want to grind your own meal and flour for table use in addition to your animal feeds, the burr mill may be the one you want. None of the whole grain is lost in a burr mill, and, generally speaking, it grinds finer (though you may have to run your meal through it twice to get the finest

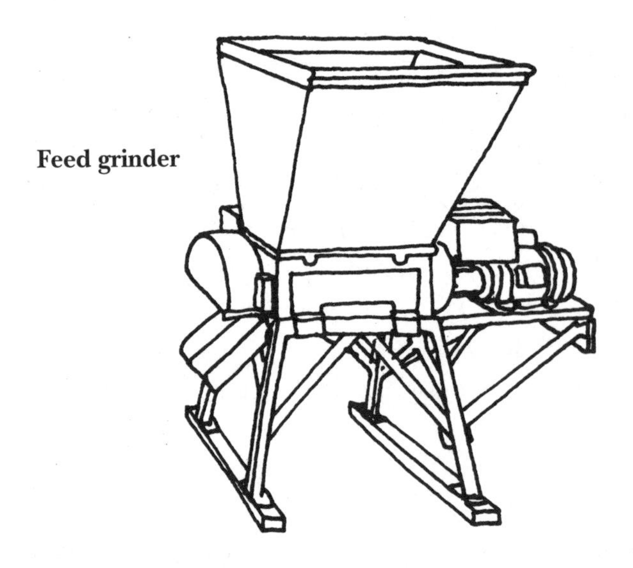
Feed grinder

flour). Old burr mills are often available too, though you can probably count on having to replace the steel burrs.

Harrell Noble, who was homesteading years and years before the idea became fashionable, told me of his adventure in restoring an old burr mill to use:

"I had an idea that the mill could solve at least two problems connected with home-produced flour," he told me. "It could grind a larger volume of grain than the smaller mills on the market, and could grind (I hoped) finer than the old stone gristmill burrs."

"The mill I had in the barn my grandfather had used for grinding livestock feed. It is a burr-type grinder, one that operates just like the stone gristmills, but which uses steel plates rather than stone burrs for the grinding process.

"I have a four-inch belt to run from the pulley on the tractor PTO to the mill pulley. (The mill was designed to run off a 6 to 12 horsepower motor). The tractor pulley had to be lined up perfectly with the mill pulley or the belt would run right off the pulley. This lining up requires a bit of jockeying the tractor around. Since the belt exerts considerable pull on the mill pulley, I had to stake the mill solidly with steel fence posts so it wouldn't budge during the grinding operation."

He continued: "Rather than try to grind the grain real fine in just one pass through the burrs and risk breaking something, I ground the grain three times, each time setting the grinding plates closer together. There's a crank on the mill to adjust the space between the plates.

"I poured shelled corn in the hopper, opened the gate valve that let the grain flow into the burrs, and we were in business. After the first pass, the cornmeal was rather coarse, but after the third grinding it was very fine and made absolutely delicious corn bread.

"Next we tried some wheat. Again I had to run the flour through the mill three times to get it as fine as I wanted. My daughter immediately made biscuits and did not need to add one bit of white flour for lightness. The biscuits were out of this world. Next day, my wife made yeast rolls, and that's when I knew for sure all the work had been worth it."

Noble's burr mill is a model made between 1909 and 1948. Burr-type mills are still made, however. The first farm-implement dealer I called had two used ones (about $300) for sale for which new parts were available. And I was assured new ones could still be purchased. Newer models are portable, that is, on wheels so they can be moved handily to where you need them. Many, if not all, newer portable mills are powered directly off the PTO shaft rather than by pulley and belt, and are therefore easier to set up for grinding. Whether any of the newer mills will grind fine flour for table use, no one I've talked to seems to know, but most figure that if you proceed like Noble has done, you should get similar results.

For animal feed, the burr mills do a good job on shelled corn and small grains, but will not grind hay or other roughages. The burrs should not be run empty. They wear out fast as it is.

HAMMER MILL

The hammer mill is most often used today to grind livestock and poultry feed. It grinds faster than a burr mill, but requires more power and is more expensive. Instead of having burrs for the grinding process, the hammer mill employs rows of free-swinging steel flails whirling at high speeds, more or less the same action as a leaf shredder. The screen under the hammers controls the fineness of the ground feed; the finer the mesh, the finer the feed. Hammer mills will grind all manner of feed, even mixed grains, ear corn, and hay.

Noble has a hammer mill also. Will it grind grain for table use?

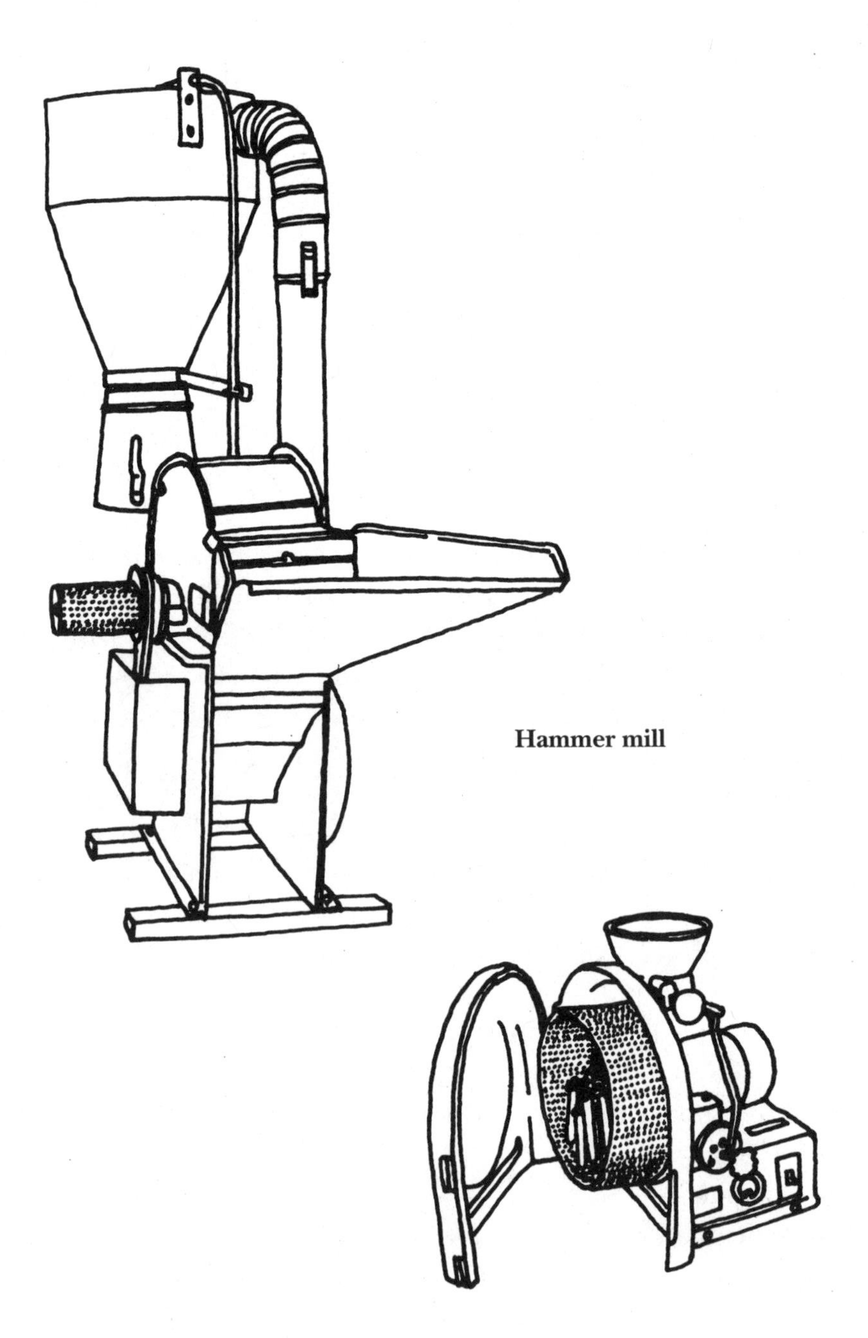

Hammer mill

"I tried several years ago to grind cornmeal in my hammer mill," he says. "In two passes, with an ⅛-inch mesh screen, I got fairly good meal, though a little coarse. With a finer screen I feel sure a hammer mill would make good cornmeal. However, one of the drawbacks of the hammer mill is that it does not produce a feed of uniform texture. Some rather coarser material is always combined with some dusty material. So you could make usable flours from all grains, but I doubt you'd ever get the truly high-quality, fine whole wheat flour for bread that you can get from a burr mill.

"For the animals though, a hammer mill is just the ticket. I use a ¼-inch mesh screen for grinding hog feed, usually shelled corn, oats, and alfalfa hay. I mix the grains beforehand, because I don't have a power mixer to mix the feed after grinding. For cattle I grind corn on the cob with oats, using the ½-inch mesh screen."

Roller Mill

Roller mills combine many attractive features for homesteaders. First of all, they are less expensive to buy, and require less power to run, than either burr or hammer mills. Grain is mashed between two rollers running with the same kind of action as the old clothes wringer. The roller surface is serrated, however, or cogged, so that the "teeth" on one roller fit the space between the teeth on the other. Grain can be merely hulled, or cracked, or ground quite fine. Some people say it's not fine enough for table use, but plenty fine for animals. Better makes (and don't ask me to give brand names here, because I would only be going on some other opinion, not having tried to grind grain for table use in a roller mill), champions of the roller mill say, will produce a fine enough flour. Be that as it may, roller mills typically lose a little of the germ, I

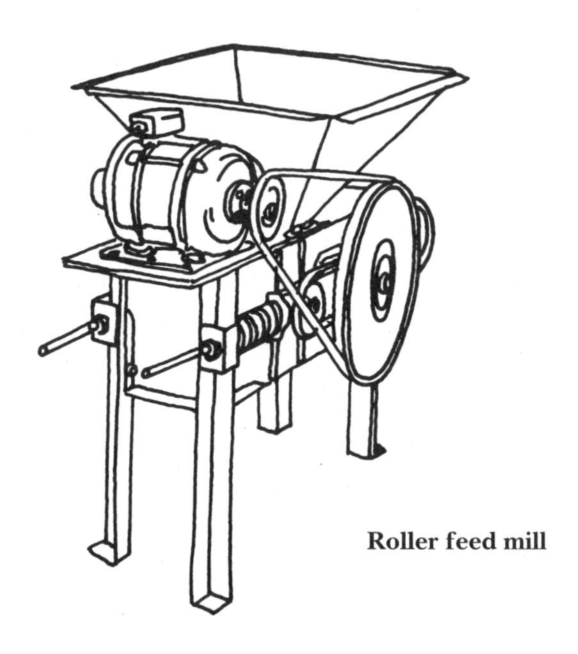

Roller feed mill

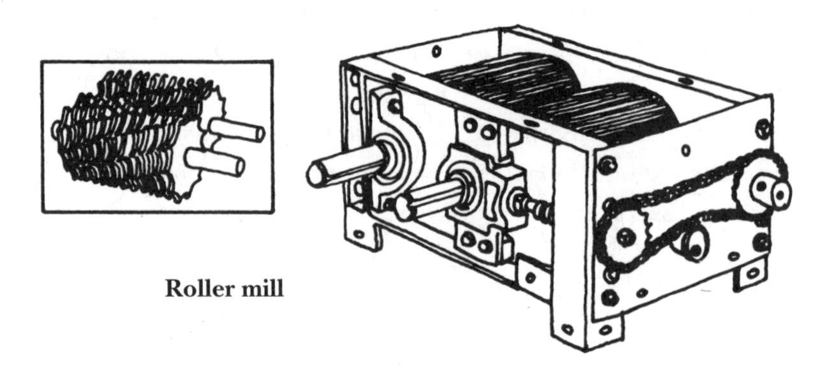

Roller mill

am told. In fact, that is one reason why roller-type mills are used in commercial flour-making: they can be set to throw out all the germ or part of it. But by the same token, you, with your own little mill, can save that germ and feed it on back through. Well worth the extra effort if you need a dehuller, cracker, crusher, and grinder all in one machine. Also, with special crusher rollers, good commercial roller mills will chop green corn, barley, or whatever into silage, and will handle any kind of grain, wet or dry.

Household Gristmills

Hand- or electric-powered mills with either steel or stone burrs, capable of turning out 5 to 50 or more pounds of flour per hour, depending on size, are available for household use. Hand-cranked models are best for producing a cup or two of freshly ground grain as needed. Stone grinders are generally recognized as grinding a finer flour than steel ones.

Be sure you know what you are buying and are buying what you want. A stone mill won't work long for grinding peanuts, or soybeans, or any high-oil kernel. Also, some mills are designed only to grind wheat and similar-sized grain and will not accept a kernel as large as corn or as small as sorghum. Nor will grain with excess moisture (over 13 percent) grind well on a stone. On the other hand, good stone grinders do not generate heat, which can ruin nutritional value.

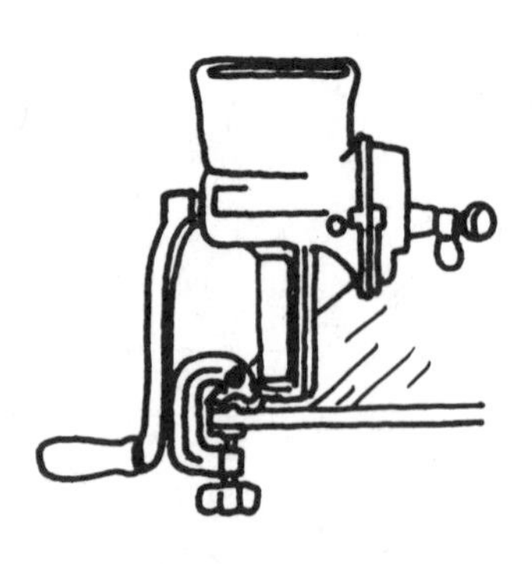

Household gristmill

Sprouting

Germinating grain and bean seeds makes them more nutritious eating. Wheat, alfalfa, triticale, soybeans, rye, barley, oats, mung beans, lentils, garbanzos, and other seeds and beans may be sprouted, the sprouts eaten as a salad, cooked into casseroles, or used in many other ways.

Devotees of sprouted grain like to refer to what they call an "explosion of vitamins" that occurs when grain sprouts. Vitamin C content of soybeans, for instance, is known to increase as much as 500 percent during sprouting. Vitamin B2 content in oats increases over 1,000 percent in five days. Increases in vitamins A, E, and K also occur. In wheat, lysine content can double. When farmers used to feed sprouted grain to their chickens, no fowl ever had it so good. Unfortunately, it rarely occurred to American farmers to try the sprouts themselves.

Sprouters

Any receptacle or set of equipment used to sprout grains for human consumption can be called a sprouter. Basically, all that such equipment must do is hold the grain in a moist condition, but not so moist as to cause mold to form. The simplest sprouter is a wide-mouth jar (or an earthenware bowl). After making sure

Sprouter

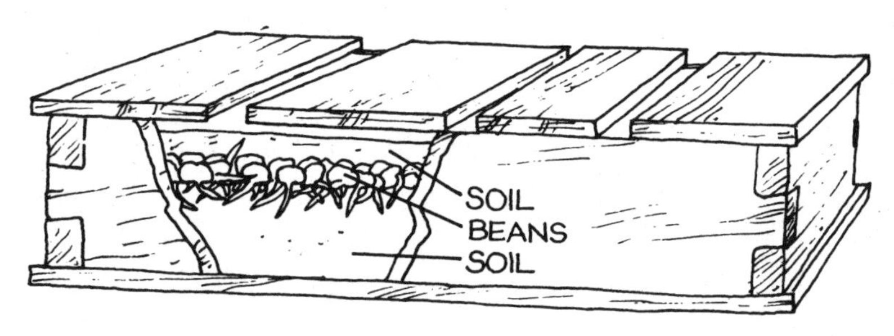

Sprouter

your seeds are not contaminated with chemical treatments for field planting, allow them to soak overnight (twelve hours) in the jar or bowl with a piece of cheesecloth, galvanized wire screen, or stainless steel screen over the top. Then pour off the water in the morning and add new water. Keep the jar in a dark place at room temperature, and rinse the seeds several times a day. Stand the jar upside down, so the water can keep draining off the seeds between rinsings. The rinsings keep mold from forming. In about four days for beans, two or three for wheat, or about six for alfalfa, you should have succulent, tender sprouts to eat.

Earth Box Sprouter

The Chinese earth box method of sprouting avoids the need to keep rinsing the seeds. Presoaked seeds are placed in a wooden box partly buried in the earth. The bottom of the box is first covered with a layer of fine soil—2 or 3 inches is enough. Then the beans are spread on the soil, but no more than two beans deep. Next another layer of soil covers the seeds. A cover on the box can be adjusted for more or less ventilation, to provide just enough circulation to prevent molds from forming. No rinsing, but of course, when you harvest the sprouts, they will need washing.

Automatic Commercial Sprouters

A fair number of sprouting devices are available commercially. Some reduce the number of rinsings necessary, and some dispense with rinsing altogether. The latter kind use capillary action to draw water up to seeds. Seeds sit on strips of special cardboard, and are kept moist, as if they were lying on a blotter.

The fault of all these sprouters, if they have a fault at all, is that only comparatively small amounts of sprouts can be produced at one time.

For much more detailed information on sprouting, check into the many resources that are available in print or online. One recently published book that discusses sprouting at home is *Fresh Food from Small Spaces* by R. J. Ruppenthal (Chelsea Green, 2008).

Index

D

Q

R

S

Y

the politics and practice of sustainable living

CHELSEA GREEN PUBLISHING

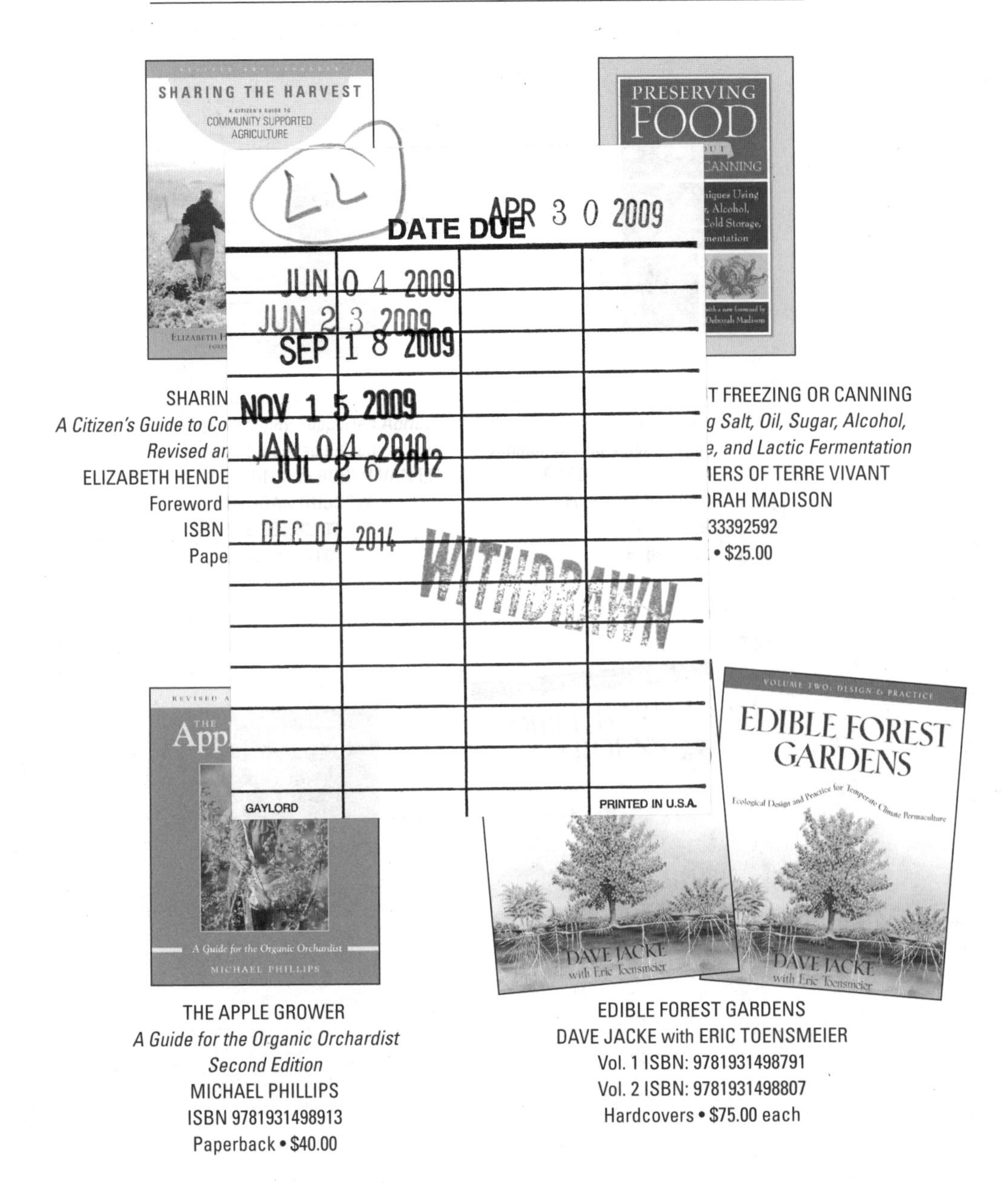

For more information or to request a catalog, visit **www.chelseagreen.com** or call toll-free **(800) 639-4099**.